Pollination Ecology

The Practical Approach Series

SERIES EDITORS

D. RICKWOOD
Department of Biology, University of Essex
Wivenhoe Park, Colchester, Essex CO4 3SQ, UK

B. D. HAMES
Department of Biochemistry and Molecular Biology, University of Leeds
Leeds LS2 9JT, UK

Affinity Chromatography
Anaerobic Microbiology
Animal Cell Culture
   (2nd Edition)
Animal Virus Pathogenesis
Antibodies I and II
Biochemical Toxicology
Biological Membranes
Biomechanics—Materials
Biomechanics—Structures and
   Systems
Biosensors
Carbohydrate Analysis
Cell–Cell Interactions
Cell Growth and Division
Cellular Calcium
Cellular Neurobiology
Centrifugation (2nd Edition)
Clinical Immunology
Computers in Microbiology
Crystallization of Nucleic Acids
   and Proteins
Cytokines
The Cytoskeleton

Diagnostic Molecular Pathology
   I and II
Directed Mutagenesis
DNA Cloning I, II, and III
Drosophila
Electron Microscopy in Biology
Electron Microscopy in
   Molecular Biology
Enzyme Assays
Essential Molecular Biology
   I and II
Experimental Neuroanatomy
Fermentation
Flow Cytometry
Gel Electrophoresis of Nucleic
   Acids (2nd Edition)
Gel Electrophoresis of Proteins
   (2nd Edition)
Genome Analysis
Haemopoiesis
HPLC of Macromolecules
HPLC of Small Molecules
Human Cytogenetics I and II
   (2nd Edition)

iii

Pollination Ecology

A Practical Approach

A. DAFNI

Institute of Evolution
Haifa University
Haifa, Israel

at
OXFORD UNIVERSITY PRESS
Oxford New York Tokyo

Oxford University Press, Walton Street, Oxford OX2 6DP

Oxford New York Toronto
Delhi Bombay Calcutta Madras Karachi
Kuala Lumpur Singapore Hong Kong Tokyo
Melbourne Auckland Madrid
and associated companies in
Berlin Ibadan

Published in the United States
by Oxford University Press Inc., New York

A catalogue record for this book is available from the British Library

Library of Congress Cataloging in Publication Data
Dafni, Amots.
Pollination ecology: a practical approach/A. Dafni.
(The Practical approach series)
Includes bibliographical references and index.
1. Pollination. 2. Botany—Ecology. I. Title. II. Series.
QK926.D213   1992   582.13'041662—dc20   92–15944
ISBN 0–19–963299–5 (hbk.)
ISBN 0–19–963298–7 (pbk.)

Typeset by Cambrian Typesetters, Frimley, Surrey
Printed by Information Press Ltd., Eynsham, Oxon.

Preface

The first evidence of Man's involvement in pollination comes from the Middle East.

The sycomore tree (*Ficus sycomorus*) is common in the Middle East but its original pollinators are not present, and consequently seeds are not produced and the syconia remain small and undeveloped.

The occupation of the prophet Amos (eighth century BC) is described in the Bible (Amos 7:14): 'I was no prophet, neither was I a prophet's son; but I am a herdsman and gatherer of sycomore fruit'. Careful examination of the Hebrew origin shows that Amos was a 'piercer of sycomores'. Keimer (1) and Galil (2) identify it with the old technique of gashing figs which is practised even today in the region in the absence of the natural pollinators. The primary purpose of gashing is to ripen the figs in as short time as possible by inducing parthenocarpic fruits. This technique was known to the ancient Egyptians for at least three millenia and special knives for this purpose have been found in tombs (1, 2).

An Assyrian bas-relief from the days of King Asurbanipal (884–860 BC) shows two deities holding a male inflorescence of date palm and engaged in the artificial pollination of the female (3, p. 9).

Several observations and conclusions concerning pollination have been made through the ages, especially in the last two centuries (4, pp. 19–32; 5, 6, 7). The founder of the modern pollination ecology is undoubtedly C. K. Sprengel who in 1793 published a pioneering book entitled *Das entdeckte Geheimniss der Natur im Bau und in der Befruchtung der Blumen* (The secrets of Nature revealed in the structure and fertilization of flowers).

Baker (7) surveyed the history of sex in flowers and pollination and divided it into three eras; the Old Testament of anthecology, which was written at the end of the nineteenth century; the New Testament, which began to be assembled after the rise of the Synthetic Theory of Evolution in the 1940s; and the Apocrypha writings, which are to be found in all periods of literature and includes additional observations, some of which have been contributed by amateurs.

Feinsinger (8) stated that only in the last decade has pollination ecology approached the status of a 'hard' science with an established theoretical framework and an established research protocol. He also realized that pollination ecology is rapidly being integrated into a balanced perspective of plant life history phenomena, plant demography, and other areas of plant–animal interaction.

Although the literature on pollination ecology is voluminous and continually and rapidly increasing, only very few publications have been devoted to the methodology of research (e.g. 9, 10).

Macior (11) was correct when he stated that

Our present knowledge of visual, olfactory, and taste perception of pollinators must be related experimentally to the structure and function of pollination mechanisms. Circumstantial evidence between spectral reflectance from corollas and insect vision, between chemical composition of nectar and pollinators' taste perception, and other similar correlations, is valuable as a starting point for testing the significance of such putative adaptations. More precise observational techniques of pollinator behaviour are also required to document the operation of contemporary pollination mechanisms and the need to avoid the observational errors of the past.

Cox and Knox (12) suggested applying Koch's postulates (13) to pollination ecology (as Cox's postulates) as follows:

1. Pollen transfer from anther to vector must be demonstrated.

2. Pollen transport by vector must be demonstrated.

3. Pollen transfer from vector to stigma must be demonstrated.

4. Pollen deposited by vector must be shown to affect fertilization.

If we adopt this methodology into a general framework, including also other aspects of pollination ecology, we may add more postulates such as:

5. Flower advertisement (visual as well as olfactory) must be demonstrated to be perceived and used by the pollinating vector.

6. Flower reward must be demonstrated to be consumed (or used) by the vector as an integral part of the pollination process.

7. The relative contribution of pollen and ovules to the next generation has to be demonstrated as a result of the pollination process.

8. Interrelationships between different vectors involved in pollination have to be demonstrated at a community-based level.

The present book is an attempt to provide a manual of the methods and procedures used today in pollination ecology. The general trend is deliberately biased toward field studies, with a special emphasis on techniques that are applicable in remote places by using the minimum possible equipment.

Many concepts in pollination research are the subject of conflicting views (e.g. flower constancy and pollination efficiency), and hence the theoretical framework of these issues has become somewhat enlarged. As one's conceptual approach may influence the methods used in a particular piece of research, an overview of commonly-used procedures is supplied where it is impossible to provide a detailed working protocol.

Some of the material presented here was generously provided by the late Irene Baker to whom this book is dedicated. Several sections were contributed by Theodora Petanidou, Heidi Dobson, and Peter Feinsinger. Other valuable information and good advice was supplied by Peter Bernhardt, Bernard Vaissiére, Andreas Erhardt, Peter Kevan, Rani Kasher, Simon Laughlin, Susanne Koptur and Avi Shmida.

The manuscript has greatly benefited by the thoughtful comments and criticism provided by Juan Arroyo, Manja Kwak, Peter Feinsinger, Adrian Meeuse, Sally Corbet, Peter Kevan, Peter Bernhardt, Christian Westerkamp, Clara Heyn, Andrew Lack, Jens Olesen, Christopher O'Toole, Dan Eisikowitch, and Robert Bertin. To all these friends and colleagues who provided good advice, criticism, information, and a lot of encouragement, I want to express my deepest feelings of gratitude.

References

1. Keimer, L. (1927). *Biblica*, **8**, 441.
2. Galil, J. (1967). *Econ. Bot.*, **22**, 178.
3. Meeuse, B. J. D. (1961). *The Story of Pollination*. Ronald Press, New York.
4. Proctor, M. and Yeo, P. (1973). *The Pollination of Flowers*. Collins, London.
5. Schmid, R. (1975). *The Biologist*, **75**, 26.
6. Lorch, J. (1978). *Isis*, **69**, 514.
7. Baker, H. G. (1979). *N.Z. Jour. Bot.*, **17**, 431.
8. Feinsinger, P. (1987). *Rev. Chile. Hist. Nat.*, **60**, 285.
9. Porsch, O. (1922). Methodik der Blütenbiologie, Abderhaldens Handbuch. *Biol. Arbeitsmeth.* Abt. X, Teil 1, Heft 4, Lief. 81. Wien.
10. Meeuse, B. J. D. (1977). *Reproductive Biology of Flowering Plants*. ASUW Lecture Notes, Botany 475, 113 HUB, FK-10, University of Seattle.
11. Macior, W. L. (1974). *Ann. Missouri Bot. Gard.*, **61**, 760.
12. Cox, P. A. and Knox, R. B. (1988). *Ann. Missouri Bot. Gard.*, **75**, 811.
13. Koch, R. (1880). *Investigations into the Etiology of Traumatic Infectious Diseases*, (trans. W. W. Cheyne). The Sydenham Society, London.

Haifa and Cambridge   A. D.
January 1992

In memory of Irene Baker,
a generous and dedicated pioneer
in experimental pollination biology.

Contents

Contents

Contents

Abbreviations

AD	aniline-diphenylamine
AVS	animal visual spectrum
FCR	fluorochromatic reaction
GC	gas chromatography
GC-MS	gas chromatography/mass spectrometry
GSI	gametophytic self-incompatibility
HVS	human visual spectrum
IPT	improper pollen transfer
ISI	index of self-incompatibility
IVS	insect visual spectrum
NAD	nicotinamide adenine dinucleotide
OCI	outcrossing index
RH	relative humidity
P:O	pollen:ovule
RF	rates of flow
SEM	scanning electron microscope
SI	self-incompatibility
SSI	sporophytic self-incompatibility
TLC	thin-layer chromatography
UV	ultraviolet

1

Functional floral morphology and phenology

1. Flower structure in relation to pollination

Angiosperm flowers are very diverse in size, shape, and colour. Functionally, a flower is a compound organ in which all its structural complexities are presumably adapted to sexual reproduction. Therefore, flower structure, phenology, and the evolutionary ecology of pollination partnerships are interwoven, so much so that systematists rely on floral structure for identification and phylogenetic studies (1).

Each part of the flower may have a special role in one or more events during production and dispersal of gametes and seeds. The organization of inflorescences, the spatial positioning of the blossoms (*sensu* Faegri and van der Pijl (2), who defined blossom as 'the pollination unit as an ecological term to differ from flower which is a morphological term'), and their position on the plants, are regarded as adaptations which increase the chances of pollination.

It is usually assumed that every floral organ has a more or less definite role in pollination, but replacement of functions is common (see *Table 1*). A typical hermaphrodite flower of the angiosperms comprises a repetitive pattern which consists of sepals, petals (or tepals), stamens, and carpels. The outer whorls serve mainly in protection and advertisement. The reward-producing organs are almost always sporophytic structures. The gametophytic structures develop within the carpel and the anther (the egg cell and the pollen grain, respectively). Sometimes, floral sexuality departs from the hermaphrodite pattern to produce various sexual types among flowers, individual plants, and populations (see Chapter 2, *Table 1*). The dichotomy between asexual functions (protection, advertisement, and reward) and sexual functions (producing gametophytes) is not strict—anthers and carpels may have infertile parts which contribute to the process of pollination beyond their immediate sexual role (see *Table 1*).

Table 1. Floral functional morphology in relation to pollination

Organ	Function		Examples
Calyx	Protection	of the whole flower when closed of young buds prior to flowering of ovules (e.g. bird-pollinated flower) from nectar larceny	*Papaver*
	Advertisement	Visual—attraction	*Limonium* spp.
		Olfactory—glands of essential oils	Many aromatic Lamiaceae
Corolla/perianth	Reward	Nectaries—partly or as a whole	
	Advertisement	Long range { Visual: colour, position; size, inflorescences, shape, organization / Olfactory: highly dispersed volatiles	
		Close range { Visual: nectar guides / Olfactory: odour guides (osmophores) / Tactile: surface microstructure	
	Directing of pollinator behaviour	Restricting diversity of visitors { spurs / tubes / hairs, scales / proboscis guides / floral complexity	*Aquilegia* / *Salvia* / *Anchusa* / *Anacamptis*

	Function	Description	Examples
Shelter	Lodging (night or bad weather)		
Deception	Imitation	resemblance to a rewarding flower; pollen mimic	*Cephalanthera*
	Labellum	mate mimic (pseudocopulation); traps	*Ophrys*
Reward	Nectaries (glands + accumulating cavities)		*Cypripedium*
Protection	Heating (higher temperature than the surroundings)		*Serapias*
	Closing at night or in adverse weather		*Crocus, Anemone*
Androecium			
Filament:			
Protection	of nectar (e.g. nectar chamber)		*Asphodelus, Citrus*
	of the female organs by creating a tube around them		*Leguminosae*
Advertisement	Hairs—enhancement of attraction at close range		*Verbascum, Anagallis*
	Coloured and protruberating		*Capparis, Eucalyptus*
Reward	Bearing nectaries		*Colchicum, Persea*
Anther:			
Reward/reproduction	Pollen production		
Deception	Persistent, coloured after pollen shed		*Saintpaulia*
	Coloured as a fale attraction		*Canna*
	Swollen but empty, mimesis of fertile one		*Dianella, Cleome*

Table 1. *Continued*

Organ	Function		Examples
Connective:	Pollination mechanism	Sterile, clutch mechanism of pollen deposition	*Salvia*
Staminodes:	Protection	Provide shelter	Relict Magnoliales
	Advertisement	Enhance flower conspicuousness	
	Reward	Bearing nectaries	*Laurus*
			Sparmannia africana
	Deception	Edible	*Dieffenbachia longispatha*
		Pollen mimesis	
The whole anthers			
Heteranthery:	Reward/Reproduction	Arranged in two heights exploited by different pollinators	*Sternbergia clusiana*
		[Feeding anthers ($\pm$ cryptic) as a reward	
		[Exposed anther mainly for reproduction	*Viola, Swartzia*
		[Nectar gland on stamen	
Enantiomorphy:		[Mechanism to promote outcrossing, pollination efficiency, and pollination precision	*Cassia*
False anthers:	Advertisement/ Deception	[Increasing flower attraction	*Titania fugax*
		[Economy (a smaller quantity of pollen required)	
Pollinaria:		All the pollen removed as one or two units	Orchids, Asclepiads
Stylopodium	Reward	Edible, parts of	*Shorea*

Gynoecium

Stigma			Example
Pollen reception	Spatial fit-in structure to catch pollen		
	Biochemical-serological recognition		
	Germination substrate		
	Stigma closure	to provide favourable microclimate for the pollen grains	
		to prevent loss of already located pollen	Asteraceae
Breeding	Secondary pollen presentation		
	Stigma closure as a mechanism to avoid self-pollination		
	Hidden stigma and cleistogamy		
Advertisement	Showy lobes and crests		*Crocus*
Reward	Stigmatic fluid		*Petunia*, Aroids, *Asclepias*
Deception	Lobed, large and yellow stigma in ♀ flower resembles the anthers of ♂ flower (form of automimicry)		*Begonia*

Style:

		Example
Pollen reception	Spatial positioning of the stigma to face the pollinator	
Advertisement	Petaloid style	*Iris*
Directing the pollinator	Half tubular, restriction of pollinators	*Iris*

Ovary:

		Example
Reproduction	Bearing ovules	
Reward	Bearing nectaries	*Urginea*
	Oviposition substrate	*Ficus*
	Food for adult insects (beetles)	

Table 2. Main blossom types and their functional characteristics[a]

		Bowl–dish	Bell–funnel	Brush	Flag	Gullet	Trumpet	Tube	Traps
Advertisement	Visual attraction	Corolla (petals or perianth)	Corolla (petals or perianth)	Stamens	Standard	Lips	Limb and Tube	Corolla Tube	Corolla, or perianth (Tube, Spadix)
Aspects of pollinators' behaviour	Landing surface	Whole flower	Flower margin	Any part	Keel	Lower lip	Margin	Margin	Margin
	Hold fast	–	–	–	Relies on lateral petals	–	–	–	–
	Guiding marks	Often none	Often none	None	Symmetry, marks on standard	Nectar guides	± Nectar guides		Coloured pattern
	Guiding structures	Often none	Often none	None?	Floral structures as a whole	None or hairs; floral structures as a whole	Surface structure, hairs, ridges	Narrow tube, scales hair	Hairs
Pollen presentation		Exposed	Partially hidden ± centralized	Exposed	Well hidden	Well hidden	± Hidden	Well hidden	Hidden to well hidden

	± Localized	Diffuse	Centralized (if present)	Centralized	Centralized	Centralized	Centralized
Nectar offset							
Reward							
Pollen	+++	+	±	+	+	±	±	–
Nectar	–	+	+	±	+	++	+++	–
Main pollinator	Beetles non-specialized bees and flies, Lepidoptera	Short-tongued bees and wasps, flies, settling moths, birds, bats	Bees, butterflies, beetles, mammals, birds	Long-tongued bees, birds	Long-tongued bees, hawk-moths, birds, butterflies	Butterflies, hawk-moths, birds	Hovering or perching moths, butterflies, birds, long proboscis flies	Carrion flies and beetles, micro-diptera, bees (orchids)

[a] Greatly modified from ref. 2.

A basic understanding of floral structure, sexuality, and phenology is a prerequisite for understanding the floral life cycle as well as a necessary background for any pollination study. It is almost impossible to divorce the architecture of an organ from its function (*Table 2*). However, for practical reasons, advertisement and reward will be dealt with in separate chapters, while bearing in mind that they are functionally interdependent and therefore inseparable. (For descriptions and terminology of flower structure see ref. 3, or any other standard botanical textbook.)

2. Flowering phenology

Flowering phenology should be studied at six levels:

- the single flower (including variation in the sexual organs—dichogamy) (4)
- the individual plant
- dioecious plants
- the population
- the community (5)
- the phylogenetic aspects (6)

Phenological records depend on the parameters chosen by the various investigators (see *Table 3*) and depend on the research level, the aims of the research, and the type of analysis. The main events are the timing, duration, sequence, intensity, and frequency of flowering, which are determined by the physical environment (e.g. temperature, rainfall, and day-length). All are a product of mutualistic relations and selection involving other organisms (co-evolution). An attempt to summarize the main current issues in flowering phenology is presented in *Table 4*.

Events and changes in the single flower are recorded for studying the following:

(a) Flowering phenology in relation to geographical (e.g. latitudinal, altitudinal) and climatic variables.

(b) The time and duration of pollen exposure and stigma receptivity, and the interrelations between them.

(c) Flower advertisement in relation to reward, pollen exposure, pollinator activity, and behaviour; and to the pollination rate and efficiency (see Chapter 6, *Table 4*).

(d) The interrelations among environmental variables (e.g. temperature, humidity, light intensity, day-length) and flower development rate, reward presentation, pollinator activity, and reproductive success through pollination.

(e) The flowering variables in relation to pollinator abundance, competition

Table 3. Common flowering phenology parameters, their definitions and uses

Parameter	Definition	Level of organization	Remarks
Flowering commencement	The date (or the day number in the year) of the first flowering	Fl, Pl, Po, Sp, Co	Commonly correlated with geographical coordinates (altitude, latitude, and longitude along ecological gradients) and environmental variables (temperature, humidity); used to define flowering sequences and correlations with pollinators' availability and seed production
Rate of flowering	The cumulative no. of flowers vs. time Cumulative percentage of flowers in anthesis	Pl, Po, Sp Inf, Pl, Sp, Po, Co	The basic data for the definition of the species' phenological rhythm. Studied in relation to pollinator availability, pollination intensity vs. time, cross-pollination, and seed production
Course of flowering	No. of flowering units vs. time Date vs. proportion of flower-day: i.e. the number of flowers that were open that day, divided by the total number of flowers counted on all days sampled (7)	Pl, Po, Sp, Co	Used to compare the flowering yearly spread (e.g. flowering species vs. month) of various life-forms, species at different vegetation layers, or micro-habitats at the same community or along an ecological transect. Useful especially in large samples, e.g. in seasonality modelling studies
Peak of flowering	The date of maximum no. of flowers flowering plants flowering species	Pl Po, Sp Co	Commonly compared with: geographical coordinates, environmental variables, pollinator abundance and pollination rate
Relative flowering intensity	No. of flowers at the individual plant peak as % of the highest no. of flowers of its conspecific individual (8)	Po	Measure of the spatial distribution of floral resources which may influence patterns of pollen flow

Table 3. *Continued*

Parameter	Definition	Level of organization	Remarks
Index of flowering magnitude	The number of simultaneous open flowers in a given time as the percentage of the total number of flowers	Inf, Pl	Measure of magnitude of floral resources
Mean flowering duration	The mean flowering duration of the sample in days	Inf, Pl, Sp, Po, Co,	Largely used to study seasonal flowering sequences, co-flowering and over-lappings in relation to the pollinators' occurrence and abundance
	The mean flowering duration of the sample in days	Po	A comparison between the mean duration of species flowering and individuals' durations may show if the individuals of the population have a long or staggered flowering duration
Midpoint flowering time	The midpoint of extreme record dates of flowering (9)	Po, Sp, Co	Although used for comparing co-flowering species, this measure ignores the real flowering duration of each species, since the mid-date does not necessarily correspond to the peak of flowering. A useful criterion for comparing large data sets compiled from floras, etc.
Dispersion of the flowering curves during the flowering season	The observed variance of the distance between mean flowering dates is compared with the expected variance based on randomly dispersed means and a uniform distribution	Co	Studied in relation to the chances for pollination success earlier or later than the rest of the species and to test hypotheses concerning competition for pollinator availability
	If the ratio of the observed variance to the expected variance is much greater than 1, the dispersion is aggregated; if the ratio is much less than 1, the dispersion is even (10)		

Flowering overlaps

Co

Comparison of the observed flowering overlaps between each two species with those generated by a series of randomizations, as a criterion for minimization of phenological (flowering) overlap with the remaining species (11, 13)

Maximal no. of plants found simultaneously in peak blooming, expressed as a % of the sum no. of its conspecifics in the sample (8)

Co

Dividing the shared area of the two flowering curves of each species pair by the total area under the two flowering curves (12)

Degree of overlap in flowering among individuals, calculated for all possible pairs of n individuals. $C = a/b$, a = the number of census dates on which plants flowered strongly, b = the number of census dates for whichever individual flowered strongly for the fewest dates. Mean intraspecific overlap. $Z = \Sigma\ c/N$, when N is the number of comparisons. $N = n(n-1)/2$, when n = the individual number $Z = 1$ in complete overlap, and 0 with no overlap (15)

Studied in relation to species flowering sequence. The assumption is that consistent competition for pollination may result in non-random flowering phenologies leading to decreased flowering overlaps.
(See ref. 12 for review)

In random models (12):

(a) The time for flowering is subjectively defined, and different durations yield different results

(b) A uniform distribution of resources (time for flowering) is assumed

(c) Rejection of a regular dispersion pattern would not preclude specific cases of divergence

(d) In the testing of competition, only species that share pollinators should be included

(e) The dispersion of the flowering mid-point does not disclose any information regarding overlapping among species

Used to test aggregated vs. dispersed flowering phenologies in relation to competition for pollinators

Used to study variation among individual plants and as a measure of synchronization (= uniformity)

Need individual measurements of the flowering duration. Note the controversy between authors on the approaches to define and measure overlaps (14)

Table 3. *Continued*

Parameter	Definition	Level of organization	Remarks
Consistency of flowering sequence over the years	The species in the sample are ranked according to their mean flowering time each year. Ties are assigned an average value. Rank values are then correlated between years, using Spearman's rank order correlation coefficient. Kendall's coefficient of concordance is calculated to measure the degree of association among years (12)	Pl, Sp, Po, Co	See remark on mean flowering duration
Flowering termination	Last date of flowering	Pl, Po, Sp	See commencement and duration of flowering, also studied in relation to seed set success

Legend: Levels of organization

Fl = flower
Pl = individual plant
Sp = species

Inf = inflorescence
Po = population
Co = community

Table 4. The interrelationships among the timing, duration, and frequency of flowering, and the level of organization

Level of organization	Timing of flowering	Duration of flowering	Frequency of flowering
Intra-floral	Synchronous (Homogamy)	Flower longevity varied considerably among species, families, habitats, seasons, dates, and breeding systems (15)	
	Asynchronous (Dichogamy) {Female first (Protogyny)	Female phase is longer (16)	
	Male first (Protandry)}	Male phase is longer (17, 18)	
Monoecious species	Female flowers seem to open before male flowers (19)		
Dioecious species	In most species males tend to flower earlier (20)	Female flowers last longer than the males (15)	
	Early flowering of male flowers may accustom pollinators to visit these plants and thus increase their potential as pollen donors (21)	Male plants appear to flower as long as or longer than females (20)	
Andromonoecious species	Bisexual flowers open earlier than male flowers (22)	Individuals often bear male flowers over a longer period than bisexual flowers (24)	
	Various patterns in Apiaceae (23)	In Apiaceae male and bisexual flowers have roughly the same duration (25)	
Flower/blossom	± At a fixed time {Coincidence with the ± specific pollinator, better chances for a pollinating visit (26)	(a) Large flowers, higher investment, more nectar → extended floral life-span; compensation for the investment (27)	

Table 4. *Continued*

Level of organization	Timing of flowering	Duration of flowering	Frequency of flowering
Flower/blossom (cont.)	At various times during the day — Various pollinators available throughout the day At the season — Correspond to the emergence or migration of the pollinators (28)	(b) Habitat-dependent → flower longevity increases with cooler day and night temperatures (15) In tropical forests many species have one-day flowers (29) (c) Inflorescence Gradual ripening: elongation of anthesis duration Synchronous flowering: environmental constraints on long exposure and/or high frequency of pollinators	In geophytes, it may be dependent, on the accumulation of sufficient reserve level (31)
Individual plant	Asynchronous within the canopy } Reducing geitonogamy (30)		
Population/species	*Asynchronous:* (a) Reducing intraspecific competition for pollinators (30) (b) Results in the reduction of the effective population size (13) (c) Increases the diversity of matings (30) (d) Local patchy weather conditions	*Extended:* (not in masses) (a) Rate of flower production can be adjusted to resource availability needed for fruits (30) (b) Enhancement of cross-pollination towards an increasing diversity of matings (30) (c) Lack of seasonal differences in resource of pollinator availability (14) (d) As a response to environmental cues: temperature, and moisture (32) photoperiod	

(a) Stabilizing selection for outcrossing (33)

(b) Attraction of more pollinators (33)

(c) Reflection of uniform and unambiguous environmental causes (34)

Variation occurs across the geographic range of the species, due at least in part caused by variation in latitude, elevation, etc. (37–39)

Populations from higher latitudes flower earlier than populations from lower latitudes when planted at low latitudes (39, 40)

Populations from high altitudes flower later than those from lower altitudes (38, 41)

Flowering time is regulated by seed dispersal requirement (44) and seed predation (45)

Community

Rewarding species: Müllerian mimicry: Synchronous flowering to promote the common advertisement (47)

Early blooming
Escape in time of highly rewarding species from low rewarding ones (51)

Asynchronous:
(a) Less improper pollen transfer (14, 54)

(b) Less exposure to risky weather or pollinator scarcity (30)

(c) Reduction of interspecific competition in trapping foraging pollinators (35)

(d) Genetic variation between individuals (30)

(e) Different micro-sites in patchy environment (30)

Shortened. 'Multiple bang' and 'Big bang species' many flowers in a short period of several days (36)

Mass flowering: for several weeks.
(a) Attraction to overcome rarity or high dispersion in space (30)

(b) Escape in time from bud, seed, and flower consumers (42)

(c) Long distance visual cues to locate resources for pollinators (30)

(d) Species with supra-annual blooming cycles which are rare in time (43)

Extended blooming:
(a) High altitude species tend to flower for shorter periods in summer, compared to the longer blooming species which peak in winter but see ref. 46, p. 102, and ref. 37 for different results)

(b) Non-rewarding species will maintain low in flowering density during the season then flower with or after other sympatric rewarding species (48)

(c) In early successional species, prolonged production → more chances for recolonization (52)

(d) Enhancement of the visual cue (30)

Multiple bang: Many flowers are produced for short periods several times a year (36)

Monocarpy:
(a) Result of satiation of seed predators (49)

(b) Increases the chances of survival in the space vacated by the adult plants (50)

(c) Result of selection for disproportionate gain in fitness with increasing reproduction effort (53)

Table 4. *Continued*

Level of organization	Timing of flowering	Duration of flowering	Frequency of flowering
Community (*cont.*)	Simultaneous co-blooming (convergence)	*Mass flowering:*	*Supra-annual blooming cycles*
	(a) Batesian mimicry, the mimic overlaps the model flowering period (56)	(a) Enhancement of pollinator fidelity and interspecific gene flow (55)	Long time for fruit maturation (30)
	(b) Reducing the overall uncertainty of foraging in the habitat (57)	(b) Less interspecific competition associated with pollinators that remain constant to a resource patch (58)	*Polycarpy:* One flowering per year, environment with predictable annual cycle (30)
		(c) Outcrossing may be enhanced through intensive competition among pollinators at the resource patch (59)	*Several episodes per year*
			(a) Bet-hedging strategy faces uncertainty in the environment (weather, pollinators, seed dispersal) (17, 60)
	Deceptive species:	*Shortened:*	(b) When plants offer no floral reward, but bloom repeatedly over short periods after highly rewarding species (36)
	Batesian mimicry—the model penetrates the flowering period of the model (51, 61)	(a) Batesian mimicry, the mimic flowering duration is shorter than that of the model (62)	
	Non-model deception. Early blooming before the naïve pollinator establishes a rewarding search-image (61, 62, 63)	(b) As a result of selection against interspecific gene flow which may cause reduction in seed set (54)	(c) Production of a few relatively small 'clutches' rather than a single large one (30)

Sequential:

(a) Facilitation of nectarless (or less reward) species flowering after high reward ones ensures pollinator visits (36)

(b) Plant species flowering early in the season may support the initial pollinator population of later species' 'sequential mutualism' (64)

The biogeographical history of the plants and the pollinators determines the co-occurrence of both through the season (65)

(c) Reduction/avoidance of competition, especially in sympatric congeners which share pollinators (39)

Segregated during the day:

(a) Reduction of competition for the same pollinator (66)

(b) Adaptation for attracting specific pollinators (2)

Seasonal temporal segregation

Reducing interspecific gene flow (36)
Divergence in flowering time used as an indication for the evolution of competitive avoidance (39) as well as a result of fruit and seed biology (67)

Phylogenetic

Phylogenetic constraints at the family level, may be more important in the variation in flowering time than in competition (6) and/or mutualism between plant species, sexual selection in plants, competition between pollinators, and reward partitioning during the floral life-cycle.

The recording of flowering events (time of occurrence, duration, and termination) is not a goal *per se* in pollination studies unless implications or hypotheses are set forth. Thus, this recording should be related to other variables such as:

- individuals, populations and/or species
- physical parameters (e.g. temperature, day-length, altitude, latitude, and relation to a certain time)
- biological agents (pollinators or predators)
- advertisement intensity and reward structure

When recording events, quantitative analysis of the data is needed. The size of the samples practically available for study is dependent upon plant size, life form, and population density and size. Pragmatic constraints also apply. It is advisable that, when choosing the site and with regard to field marking, easy access to the growing site and the flowers should be considered so as to make the best use of time and effort. The sample sizes as suggested herein are only tentative, and should be modified in accordance with the actual situation.

The frequency of observation depends largely on the type of inquiry. It may be in intervals of several minutes or hours while dealing with the development of a single flower, or every two to four weeks if the research is focused at the community level. Frequency of observations should be chosen so as to minimize the loss of relevant information required for the specific study and the type of flowering rhythm. In large or tall plants it is useful to mark a sample of branches (say 10–20) at different locations in the canopy, and subsequently to make extrapolations for the whole plant, or when looking for differences between branches, since some autoregulation and independence has been found in large trees (J. Arroyo, pers. comm.)

In both phenological observations and examinations of the breeding system, special attention should be paid to differentiation between individuals and clones.

Protocol 1. Events in the single flower

Materials

- coloured (or numbered) tags or threads which are durable under field conditions

Method

1. Mark at least 5 flowers × 5 plants at the bud stage on different parts of the plants. Each bud should have a distinctive marking or label for further identification and observation.

2. Observe flowering progress (preferably under natural conditions) every day until flower opening, and then every other hour. Note that some flowers open rapidly, in 10–20 min from closed buds to fully expanded flowers, while others may take a few hours to open. On the day of flower opening, observations should take place throughout the day (and night, depending on flower type and pollination syndrome) at short intervals (about every 2–3 h).

3. For every observation, complete a full set of data for each flower.

4. The collected data should include:

   • plant species; details of the voucher specimens
   • locality, habitat, date, time, plant number, and flower number
   • the sex distribution (plant's sexuality) and spatial distribution of the various flower types (see Chapter 2, *Table 1*)

   • flowering stages:
     A small bud, petals not visible yet
     B large bud, petals are visible but not expanded
     C flower opening
     D full blooming (anthesis):
        D_1 before pollen exposure
        D_2 pollen exposure
        D_3 after pollen exposure
     E flower wilting

   • flower organs wilting order:
     A androecium
     B gynoecium
     C petals
     D sepals

   • position of flower relative to plant axis:
     A vertical
     B at 45 degrees
     C horizontal
     D at 135 degrees
     E at 180 degrees

   • sepal/petal/perianth development:
     A colour changes
     B size changes (diameter, expansion, elongation)

Protocol 1. *Continued*

- stamens:
  A distance of anthers from perianth
  B distance of anthers from stigma(s)
  C anther dehiscence (mode of)

- stigma:
  A diameter, colour, and shape
  B stickiness, receptivity (see Chapter 3, Section 6)
  C position (upright, bent, etc.)

- nectar production [Yes (+) or No (−)]

- odour [Yes (+) or No (−)]

5. Record the number of parts for each flower whorl, their relative arrangement (intra- and inter-whorl), their size and form, and any temporal and spatial changes in position, size, and colour of every organ.

6. Use the data to delineate the following: timing and duration of flower opening, pollen exposure, stigma receptivity, nectar and odour presence or absence; critical signs (colour, size, change of position, etc.) which indicate the onset and termination of each stage.

7. Express the percentage of open flowers presented as:
   Number of open flowers on a particular day/(open flowers + closed buds + dead or fruiting flowers).

8. Record the direction of flowering within the inflorescence (acropetal vs. basipetal or centripetal vs. centrifugal).

Protocol 2. Dynamics of flowering of the whole plant

Materials

- coloured (or numbered) tags or threads which are easily recognized and durable under field conditions

Method

1. Mark 20 plants (sample size depending on the situation) before the start of flowering.

2. Observe the flowering magnitude of ten to fifty flowers on the marked plants (every day) using the following criteria:

   - 1. before flowering
   - 2. flowering commencement (up to 25% of the flowers are open)
   - 3. before peak of flowering (25–50%)
   - 4. peak of flowering (> 50% or more open)

- 5. after the peak of flowering ($< 50\%$ blooming)
- 6. termination of flowering ($< 10\%$ open)
- 7. flowering complete (no more flowers)

Protocol 3. Flowering course of the whole population

Materials

- coloured (or numbered) tags or threads which are durable under field conditions

Method

1. Delimit the sample units before flowering commencement (for example, in transects of 100 m long or squares or circles of various sizes depending on the plant size, the density of the population, and the physical conditions).

2. Observe the flowering magnitude of the whole sample using the following criteria:
   - 1. before flowering
   - 2. to 25% of the individuals flowering
   - 3. 25 to 50% of the individuals flowering, the rest still with closed buds
   - 4. 50% or more of the individuals flowering
   - 5. 25% to 50% of the individuals flowering, the rest already withered
   - 6. less than 25% of the individuals flowering
   - 7. end of flowering (less than 10% of the individuals flowering)
   - 8. flowering complete

References

1. Kevan, P. G. (1984). In *Plant Biosystematics* (ed. W. Grant), p. 271. Academic Press, Canada.
2. Faegri, K. and van der Pijl, L. (1979). *The Principles of Pollination Ecology* (3rd edn). Pergamon Press, Oxford.
3. Weberling, F. (1989). *Morphology of Flowers and Inflorescences*. Cambridge University Press.
4. Lloyd, D. G. and Webb, C. J. (1966). *NZ J. Bot.*, **24**, 135.
5. Primack, R. B. (1985). In *The Population Structure of Vegetation* (ed. J. White), p. 571. W. Junk, Dordrecht.
6. Kochmer, J. P. and Handel, S. N. (1986). *Ecol. Monog.*, **56**, 303.
7. Rathcke, B. J. (1988). *Ecology*, **60**, 446.

8. Herrera, J. (1986). *Vegetatio*, **68**, 91.
9. Anderson, E. and Hubricht, L. (1940). *Bull. Torrey Bot. Club.*, **67**, 639.
10. Poole, R. W. and Rathcke, B. (1979). *Science (Wash.)*, **203**, 470.
11. Murray, K. G., Feinsinger, P., Busby, W. H., Linhart, Y. B. Beach, J. H., and Kinsman, S. (1987). *Ecology*, **61**, 1283.
12. Rathcke, B. J. (1986). In *Ecological Communities: Conceptual Issues and the Evidence* (ed. D. R, Strong, D. Simberloff, L. G. Abele, and A. B. Thistle), p. 383. Princeton University Press, Princeton, NJ.
13. Poole, R. W. and Rathcke, B. J. (1979). *Science (Wash.)*, **203**, 470.
14. Rathcke, B. J. (1988). *J. Ecol.*, **76**, 975.
15. Primack, R. B., (1980). *J. Ecol.*, **68**, 849.
16. Primack, R. B. (1985). *Annu. Rev. Ecol. Syst.*, **16**, 15.
17. Garnock-Jones, P. J. (1976). *NZ J. Bot.*, **14**, 291.
18. Schemske, D. W., Willson, M. F., Melampy, M. N., Miller, L. J., Verner, L., Schemske, K. M., and Best, L. B. (1978). *Ecology*, **59**, 351.
19. Stephenson, A. G. and Bertin, R. I. (1983). In *Pollination Biology* (ed. L. Real), p. 110. Academic Press, Orlando, Florida.
20. Lloyd, D. G. and Webb, C. J. (1977). *Bot. Rev.*, **43**, 177.
21. Onyekwelu, S. S., and Harper, J. L. (1979). *Nature (Lond.)*, **282**, 609.
22. Martin, F. W. (1972). *Phyton*, **29**, 127.
23. Bell, C. R. (1971). In *The Biology and Chemistry of the Umbelliferae* (ed. V. H. Heywood), p. 93. Academic Press, New York.
24. Bertin, R. I. (1982). *Ecology*, **63**, 122.
25. Lindsey, A. H. (1982). *Syst. Bot.*, **7**, 1.
26. Cruden, R. W. and Hermann, S. M. (1983). In *The Biology of Nectaries* (ed. B. Bentley and T. Elias), p. 233. Columbia University Press, New York.
27. Dafni, A. (1991). *Acta Hort.*, **28**, 340.
28. Linsley, E. G. (1958). *Hilgardia*, **27**, 543.
29. Primack, B. (1987). *Annu. Rev. Ecol. Syst.*, **18**, 43.
30. Bawa, K. S. (1983). In *Handbook of Experimental Pollination Biology* (ed. C. E. Jones and R. J. Little), p. 394, Van Nostrand Reinhold, New York.
31. Dafni, A., Cohen, D. and Noy-Meir, J. (1981). *Ann. Missouri Bot. Gard.*, **68**, 652.
32. Lang, A. (1965). In *Encyclopedia of Plant Physiology* (ed. W. Rushland), p. 1380. Springer-Verlag, Berlin.
33. Augspurger, C. K. (1980). *Evolution*, **34**, 475.
34. Schmitt, J. (1983). *Oecologia (Berl.)*, **59**, 135.
35. Janzen, D. H. (1971). *Science (Wash.)*, **171**, 203.
36. Gentry, A. H. (1974). *Biotropica*, **6**, 64.
37. Arroyo, J. (1990). *Isr. J. Bot.*, **39**, 249.
38. Arroyo, J. (1990). *Flora*, **184**, 43.
39. Waser, N. M. (1983). In *Handbook of Experimental Pollination Ecology* (ed. C. E. Jones and R. J. Little), p. 277. Van Nostrand Reinhold, New York.
40. Reader, R. J. (1982). *J. Biogeog.*, **9**, 397.
41. Herrera, C. M. (1984). *Ecol. Monog.*, **54**, 1.
42. Janzen, D. H. (1974). *Biotropica*, **6**, 69.
43. Frankie, G. W., Baker, H. G., and Opler, P. A. (1974). *J. Ecol.*, **62**, 881.
44. Lamont, B., (1985) *Bot. J. Linn. Soc.*, **90**, 145.

45. Zimmerman, M. (1988). In *Plant Reproductive Ecology: Patterns and Strategies* (ed. J. Lovett-Doust and L. Lovett-Doust), p. 157. Oxford University Press.
46. Rebelo, A. G. (1987). In *A Preliminary Synthesis of Pollination Biology in the Cape Flora* (ed. A. G. Rebelo), p. 83. South African National Scientific Programmes Report, No. 141.
47. Schemske, D. W. (1980). *Biotropica*, **12**, 169.
48. Melampy, M. N. and Hayworth, A. H. (1980). *Evolution*, 34, 1144.
49. Janzen, D. H. (1967). *Evolution*, 61, 620.
50. Foster, R. B. (1977). *Nature (Lond.)*, **268**, 624.
51. Heinrich, B. B. (1975). *Evolution*, **29**, 325.
52. Opler, P. A. (1980). *J. Ecol.*, **68**, 167.
53. Schaffer, W. M. and Gadgil, M. D. (1975). In *Ecology and Evolution of Communities* (ed. M. L. Cody and J. Diamonds), p. 142. Harvard University Press, Cambridge, Mass.
54. Waser, N. M. (1978). *Ecology*, **59**, 934.
55. Antonovics, J. and Levin, D. A. (1980). *Annu. Rev. Ecol. Syst.*, **11**, 411.
56. Dafni, A. and Ivri, Y. (1989). *Oecologia (Berl.)*, **49**, 229.
57. Real, L. (1983). In *Pollination Biology* (ed. L. Real), p. 287. Academic Press, Orlando, Florida.
58. Motten, A. F. (1986). *Ecol. Monog.*, **56**, 21.
59. Frankie, G. W. and Baker, H. G. (1974). *An. Inst. Biol. Univ. Nac. Auton. Mexico*, **45**, *Ser. Botanica* 1, 1.
60. Gentry, A. H. (1974). *Ann. Missouri Bot. Gard.*, **61**, 728.
61. Carlquist, S. (1979). *Aliso*, **9**, 411.
62. Nilsson, L. A. (1981). Pollination ecology and evolutionary processes in six species of orchids. *Abstr. Uppsala Diss. Sci.* 593, Uppsala.
63. Dafni, A. (1987). In *Orchid Biology: Reviews and Perspectives* (ed. J. Arditti), **IV**, p. 79. Cornell University Press, Ithaca, NY.
64. Waser, N. M. and Real, L. A. (1979). *Nature (Lond.)*, **218**, 670.
65. Kratochwil, A. (1988). *Entomol. Gener.*, **13**, 67.
66. Gilbert, L. E. (1975). In *Coevolution of Animals and Plants* (ed. L. E. Gilbert and P. H. Raven), p. 210. University of Texas Press, Austin, Texas.
67. Herrera, J. (1987). *Ann. Missouri Bot. Gard.*, **74**, 69.

2

Breeding systems

1. Introduction

The term 'breeding systems' includes, in its broad sense, all the aspects of sex expression in plants which affect the relative genetic contributions to the next generation of individuals within a species (1). The breeding system is used in agriculture as a tool to regulate and canalize the components of fecundity for selection purposes in cultivated plants (2) as well as in genetics (3). In pollination studies under natural conditions, knowledge of the sexual systems (*Table 1*) is an essential background for evaluation of the dependence of seed production on pollination rate and type towards the understanding of the mechanisms of gene flow within and between populations (4).

The breeding system of a given taxon may also be reflected in the floral structure, advertisement and reward, and the genetical set-up. Selfers may differ from their outcrossing relatives by a series of traits such as: fewer and smaller flowers, less scent and nectar, fewer pollen grains, close proximity between the anther and the (smaller) stigma, fewer ovules but high relative seed production (3, 5).

Today the assessment of outcrossing rates focuses mainly on the analyses of genetic markers (e.g. allozymes; refs 4 and 6). In this rapidly expanding area there have been attempts to analyse the genetic consequences of pollination. Indirect methods can be used to estimate the rate of outcrossing based on morphological analysis or P:O ratio (7).

Recently, more experimental work has been carried out to elucidate the decisive roles of pollinators in the maintenance of breeding systems, in mating patterns and sexual selection, and in gene flow (1, 4).

Traditionally, breeding systems have especially been treated in relation to mechanisms which promote outcrossing with an emphasis on morphological mechanisms which may facilitate or reduce outcrossing.

Sexual expression in plants ('sexual systems', *sensu* ref. 8; see *Table 1*) may be described on three levels: the individual flower, the individual plant, or the group of plants (2). Although, generally, the sexual system is distinct and recognizable, individuals or populations may vary in sexual expression which requires a continuous, rather than discontinuous, method of sexual system

---

Table 1. Classification of the flowering plant's sexual systems

A. Spatial arrangement of ♂ and ♀ organs:
  I. Individual plants
   1. Hermaphroditic: individual plants bear only bisexual flowers.
   2. Monoecious: individual plants bear male and female organs (flowers bisexual or unisexual ♂ and ♀).
   3. Andromonoecious: individual plants bear bisexual and male flowers (♂ flowers dominant).
   4. Gynomonecious: individual plants bear bisexual and female flowers (♀ flowers dominant).
   5. Polygamomonoecious: individual plants bear bisexual flowers, ♂ and ♀ flowers.
  II. Group of plants
   1. Dioecious: individual plants bear either ♂ or ♀ flowers.
   2. Androdioecious: individual plants bear either bisexual or ♂ flowers.
   3. Gynodioecious: individual plants bear either bisexual or ♀ flowers.
   4. Polygamodioecious (trioecious): individual plants bear either bisexual, male or female flowers.

B. Temporal or spatial isolation of ♂ and ♀ organs within a hermaphroditic flower or on the co-occurring unisexual flowers on a single individual plant (monoecious species):
  I. Protandry: pollen released from anthers before stigmas become receptive.
  II. Protogyny: stigmas become receptive before pollen is released from anthers.
  III. Herkogamy. ♂ and ♀ organs mature simultaneously, spatially isolated.

C. Biochemical recognition/rejection self-incompatibility alleles:
  I. Self-incompatibility: plants are polymorphic in respect to the presence of self-incompatibility alleles. Pollinations involving pollen and stigma sharing the same self-incompatibility alleles, including self-pollinations do not result in fruit set.
  II. Self-compatibility: all pollinations, including self-pollinations, result in fruit set.

D. Systems based on variation in style and stamen length or style dimorphisms (±) self-incompatibility:
  I. Heterostyly
   1. Distyly: Individuals have either flowers with a long style and short stamen (pin) or flowers with short style and long stamens (thrum).
   2. Tristyly: Individuals have either short-, mid-, or long-styled flowers in relation to the length of the stamens.
  II. Enantiomorphy (= Enantiostyly): Individuals have both flowers with the deflection of style either to the left or right of the floral axis.

From ref. 8, modified.

---

classification. Lloyd (9) provides a quantitative approach to represent the functional gender of a given plant. Various sexual systems may have different implications for the rate of outcrossing as well as for pollination mechanisms and pollinator behaviour.

2. Regulation of the outcrossing rate

2.1 Separation of the male and female functions

Separation of the male and female functions in flowers has traditionally (see, for example, ref. 10, p. 26) been treated as a means to avoid self-pollination.

The current experimental approach also considers the role of pollinators in the evolution of these mechanisms in relation to the breeding system (see ref. 1 for review).

2.1.1. Herkogamy

Herkogamy is the spatial separation of anthers and stigma (ref. 10, p. 27). Webb and Lloyd (11) extend this term into a broader context, which also includes heterostyly, enantiomorphy, and monoecy (*Table 2*). A detailed morphological analysis of a given flower and knowledge of its temporal pattern of development in relation to pollinator foraging behaviour are needed to determine which type of herkogamy is engaged.

2.1.2. Dichogamy

Dichogamy is the temporal separation of pollen reception and deposition. When both functions operate simultaneously, it is termed homogamy or adichogamy (11). Protandry exists where pollen is available before the stigma is receptive, while in protogyny the stigma is receptive before pollen is shed. In practice, pollen shedding can be easily recognized by the naked eye or with a magnifier ($\times$10), but stigma receptivity has to be checked experimentally or chemically (see Chapter 3, Section 6). The main implications of protandry and protogyny in relation to pollination biology are summarized in *Table 3*.

2.2 Self-incompatibility (SI) systems

Self-incompatibility is a genetically controlled system which causes rejection of self-pollen (12, 13). In sporophytic SI it is controlled by the genotype of the parent plant, not by the progeny genotypes, to distinguish from inbreeding depression as a result of self-pollination or crossing (14). Rejection of unsuitable pollen is by means of chemical recognition between the pollen grain and the stigma or style and is controlled by self-incompatibility (SI) loci.

Self-incompatibility systems are classified into two types: gametophytic self-incompatibility (GSI)—the incompatible *s*-allele present in the nucleus of the (haploid–gametophyte) pollen grain which determines for which pistil the grain is compatible; sporophytic self-incompatibility (SSI)—the pollen incompatibility reaction is determined by the genotype of the pollen-producing parent (sporophyte) (see *Table 4* for associate traits of each type).

Heterostyly is a genetic polymorphism in which plant populations are composed of two (distyly) or three (tristyly) morphs that differ in the heights at which stigmas and anthers are positioned on flowers (*Table 5*). The style–stamen polymorphism is usually accompanied by a sporophytically controlled, di-allelic SI system (12, 13, 16, 17). Heterostyly has received considerable study, especially its adaptive significance and relation to the evolution of breeding systems. (See ref. 17 for review).

Table 2. Herkogamy types in relation to morphology, pollination, and evolution

	Unordered herkogamy	Ordered herkogamy		Movement	Reciprocal herkogamy	Interfloral herkogamy
		Approach	Reversed			
Definition	Pollen and stigmas are separated, pollinator often contacts both several time in unordered sequence	Pollen and stigma are positioned or moved into position along the pathway of the pollinator which is seeking reward		Movement of the floral parts is initiated or caused by the pollinator	Pollen and stigma are separated in space within each blossom, but pollination precision is maintained between blossoms or plants by the reciprocal morphs	Plants that produce both pollen and ovules in part or in whole in unisexual blossoms
		Stigma contacts first as the visitor enters and pollen is picked up later	Anthers are placed near the throat and the stigma deep within a tube		*Heterostyly* the floral morphs (2–3) occur on separate plants	
					Enantiomorphy plants have left- and right-handed flowers	
Morphological aspects	± associated with dish and brush blossoms	Require complex floral structure ± associated with few stamens and a single stigma/style. Stigmatic area often unifacial oriented outwards. Common in bell, gullet, or tube flowers	Narrow tubular flowers. Pollinator is positioned to be in contact with the stigma	Requires complex floral structure. Flower has sensitive movable parts, e.g. sensitive stigmas which close on being touched or receiving pollen. Movement of pollinia in many orchids	Most are tubular flowers	Pollen flower only; no nectar

			Reduced self-interference		
Pollinator's role	Precise behaviour is not required, unspecialized pollinators often poly-philic	Demands a great level of pollinator precision, appropriate behaviour and size often oligophilic. Efficient pollen transfer, ± effective in preventing self-interference		Particular pollinator behaviour is required with precise position	Requires appropriate size and behaviour of the pollinator	Associated with specialized pollination
Selective factors and evolutionary constraints		Selection for decreasing interference and increasing cross-pollination, promoting pollination efficiency		Selection for decreasing interference and increasing cross-pollination, promoting pollination efficiency Associated with precised pollination often in the whole genus, indicating one origin	Heterostyly does not occur in the most primitive or most specialist angiosperms (16)	

Based on ref. 11, with some additions.

Table 3. The main implications of protandry and protogyny in relation to pollination

Aspect	Protandry	Protogyny
Frequency	Widespread (Asteraceae, Campanulaceae, Lamiaceae Malvaceae, Fabaceae)	Less common (Brassicaceae, Rosaceae Araceae, Araliaceae, Cyperaceae)
Pollination mode	No special mode	Most trap blossoms. Frequent in hermaphrodite wind-pollinated species and monoecious species
Pollinator	Mainly bees and or flies	Mainly beetles and wasps
Flowering sequence in the inflorescence	In vertical acropetalic inflorescences, nectar-collecting pollinators work upwards	In vertical basipetalic inflorescences, the pollinator works downwards (few data)
Cross-pollination	Protogyny is more effective than protandry in preventing self-pollination	
Reward	Lower flowers in vertical inflorescences are more rewarding	Upper flowers in vertical inflorescences are more rewarding (few data)

Sources: refs 1, 10, 11 and R. Bertin (pers. comm.).

Table 4. Main characteristics of self-incompatibility (SI) systems

Character	Sporophytic (SSI)	Gametophytic (GSI)
Genetic control	Mostly one locus di-allelic (multi-allelic in Asteraceae and Brassicaceae)	Mostly one-locus multi-allelic
Site of pollen arrest	At stigma's surface	In style
Stigmatic fluid	Scarce ('dry' stigma: not Poaceae)	Copious ('wet' stigma)
Protein pellicles (on the stigma)	+	−
Pollen:	Tricellular	Bicellular
Floral morphology	Heteromorphic in di-allelic systems. Homomorphic in multi-allelic systems	Homomorphic

Sources: refs 3, 12, 13, and 15.

Table 5. Main characteristics of distylous and tristylous species

Characteristic	Distyly	Tristyly
Taxonomic distribution	In 24 families, e.g. Rubiaceae, Plumbaginaceae	Only in three families (Oxalidaceae, Lythraceae, and Pontederiaceae)
Morphological types	(a) Long-style with short stamens (pin) (b) Short-style with long stamens (thrum)	(a) Short-styled, anthers are at two levels (intermediate and high above the stigma) (b) Medium-length styled, and stamens above and below the stigma (c) Long-styled anthers at two levels below the stigma
Genetic control	One supergene that segregates as a simple Mendelian factor	Two loci (S, M); short styles are Ss, ss. Homozygotes are largely epistatic to M (ssM) being mid-styled form, and ssmm are long-styled morphs
Stigma	Stigmas are larger and flatter in thrums, longer cells in stylar tissues of pins	All three morphs may differ in papillae length
Pollen	Pin flowers have smaller and more pollen grains	Shortest stamens produce the smallest pollen grains, intermediate pollen grains from mid-stamen, and largest pollen grains from the largest stamens
Successful pollination	Occurs when pollen reach the receptive stigma of the other morph	Occurs when pollen from a different floral morph lands on the receptive stigma of one of the two other morphs

Sources: refs 3, 16, and 17.

Under field conditions heteromorphic SSI (heterostyly) can easily be detected (*Table 5*) while homomorphic SSI and GSI recognition require an experimental schedule of artificial pollinations and monitoring of the seed set. Estimation of the extent of SI can be achieved by comparing the seed yield resulting from cross-pollination to that from self-pollination (*Protocol 1*).

Knowledge of the SI system is required for an understanding of:

- evolution of breeding systems
- outcrossing vs. inbreeding rates
- pollination efficiency with regard to forager's behaviour
- the adaptive value of a breeding system
- gene flow and neighborhood size

3. Experimental investigation of the breeding system

The following are important in studying breeding systems:

(a) High seed production (above 80% of the flowers) under natural conditions is often associated with self-compatibility and self-pollination.

(b) Many small, closed (e.g. cleistogamous), scentless, almost nectarless, and drab flowers are frequently self-pollinated and/or agamospermous (anemophilous flowers are excluded).

(c) Because self-incompatibility systems may be maintained by the co-existence of many alleles of several genotypes, it is highly advisable to include several plants or ramets from different plants in each experiment to prevent biased results from dealing with a single genotype.

(d) The breeding system (e.g. the rate of selfing) may vary along geographical and/or ecological gradients especially under stressful conditions (e.g. deep shade, aridity, salt spray, daily rainfall, strong winds). Results from one population cannot always be applied to the whole species.

(e) Pollinated flowers (especially when arranged in dense and long inflorescences) present on the same plants should be randomized before the procedures are carried out to prevent position effects; e.g. non-random fruit production with respect to flower position. In many open inflorescences flowers nearest to the apex fail to seed.

(f) Only a small fraction ($c. \pm 20\%$ or less) of the total number of flowers of each plant/inflorescence should be used. This will prevent abortion of fertilized ovules beyond the plant's available resources to mature fruits with viable seeds. Full hand pollination (of all the flowers on the plant or inflorescence) should be carried out as a control. This is different, of course, from attempts to identify factors limiting seed production, in which pollination of all or most flowers may be important.

(g) In plants which bear enough flowers, all manipulations and controls should be on the same plant to minimize the effects of intraspecific genetic variability.

(h) Pollination bags (*Table 6*) or nets must be made of appropriate material so as to minimize any side-effects, such as reduction of light, build-up of humidity or heat, infestation by fungi or small insects, interference with flower anthesis, etc. Check frequently (especially in hot weather above 22 °C) the temperature and the humidity inside the pollination bags in relation to the ambient conditions.

(i) While checking the pollination rate under exclusion bags or nets, special attention should be given to the possibility of wind pollination by

Table 6. Pollination bags and their characteristics

Material	Advantages	Disadvantages
Fine mesh nets, bride tulle, etc.	Cheap, suitable for multiple use. Less vulnerable to micro-climatic effect inside the bag. Durable in any weather. Different mesh sizes are available	Not pollen-proof. Not water-proof
Paper	Cheap and disposable. Pollen-proof	Not durable under wet weather. Not clear, may reduce net photosynthesis and reduce seed production
Paper and propylene	Cheap and disposable. Clear (one side). Pollen-proof	Not durable under wet weather conditions
Cellophane and glassine	Cheap and disposable. Clear. Pollen-proof	Not durable
Polyester	For multiple use. Durable under wet weather conditions. Pollen-proof. Tents are suitable to cover whole plants	Expensive, not clear. Vulnerable to micro-climatic effect especially in tropical and sub-tropical conditions

airborne pollen. Several typically animal pollinated species are also pollinated to some extent by wind (see Chapter 7, Section 1).

(j) Growing of plants under an insect-excluding net has several advantages over the bagging of individual flowers:

- Less effort is invested in covering whole plants or whole shoots/branches than of individual flowers, and larger samples can be more readily obtained.
- Bagging may affect the rate of effective pollination success and reduce the rate of seed set.
- Plants can be transplanted from natural habitats prior to anthesis (provided there are adequate numbers of pollinators).
- Plants may be grown from seeds of different origin, but should be tested simultaneously.

(k) It is possible to carry out pollen transfer without emasculation if sterile male plants are available.

(l) Data from emasculation treatments must be interpreted with care:

- It does not distinguish between xenogamy and geitonogamy if the flowers are exposed to free pollination (18).

- If pollinators visit mainly for pollen they may avoid emasculated flowers, thus providing underestimates of insect pollination (18); but see ref. 19.
- The stigma and the style of emasculated flowers may function abnormally (20, 21).

(m) Pollen:ovule ratio studies (*Table 9*) should be used as a cross-reference to the results of experimental crossings.

Protocol 1. Pollination rates under natural conditions vs. spontaneous self-pollination

Materials

- numbered weather-resistant tags
- exclusion bags ('pollination bags') or fine mesh transparent nets

Method

1. Choose and tag the plants and flowers before flowering commences.
2. (a) Pollination rate under natural conditions: Leave the marked chosen plants or flowers untouched and open to free pollination.
   (b) Spontaneous self-pollination: Cover the tagged flowers or the whole plants with the exclusion bags or fine mesh before flowering, and remove any flowers already open. Emasculate a parallel sample.
3. After treatments are completed cover the flowers again and do not touch them during the observation period.
4. At the end of the season count the fruits and seeds produced in relation to the number of flowers observed.

Protocol 2. Experimental cross- and self-pollination

Materials

- numbered weather-resistant tags
- exclusion bags ('pollination bags'), or fine mesh transparent nets

Method

1. Cover the plants or flowers with exclusion bags or fine mesh nets before flowering commences, and remove any already open flower.
2. Emasculate several flowers as they open by use of a razor or fine tweezers to excise anthers before pollen is liberated, and re-cover the flower or plant.

3. Pollinate several untreated un-emasculated flowers with their own pollen and re-cover the flower or plant.

4. Pollinate the stigma of emasculated flowers with pollen from other flowers of the same plant, and pollinate a second sample with pollen from another plant. Repeat the double procedure on several plants.

5. At the end of the season count the fruits and/or seeds produced in each treatment and in relation to the number of the flowers pollinated in the same treatment.

The last two procedures (*Protocols 1* and *2*) cover only a part of the possible variations of plant breeding, and a full scheme of control and other crossings has to be made (*Table 7*) to determine the success of self- vs. cross-pollination. Additional examinations (*Table 8*) are needed to understand the role of the breeding system in the context of reproductive success.

3.1 Estimation of self-incompatibility rate

Self-pollination studies reflect fruit and seed set as a result of artificial self-pollination in the absence of foreign pollen. The studies of outcrossing rates reflect seed production under natural conditions which expose the stigma to foreign pollination as well as to geitonogamous and self-pollination. The results from the various tests should be used to define several criteria for quantitative evaluation of the breeding system.

Self-pollination rate is an important criterion to evaluate the fate of pollen on the stigma of the same flower plant or genets. High self-compatibility (which is genetically controlled) has to be treated separately from the chances of self-pollination (which is controlled by the timing of events in the flower, its structure, and pollinator's behaviour). Plants could be highly self-compatible but with low chances for self-pollination due to herkogamy (Section 2.1.1), dichogamy (Section 2.1.2), or pollinator behaviour (e.g. visiting very few flowers per plant). Zapata and Arroyo (22) suggested the following index to measure self-incompatibility (ISI):

$$\text{ISI} = \frac{\text{fruit set from self-pollination}}{\text{fruit set from cross-pollination}}.$$

The values of the ISI reflect the following possibilities:

$$>1 = \text{self-incompatible.}$$
$$>0.2 <1 = \text{partially selfincompatible}$$
$$<0.2 = \text{mostly self-incompatible}$$
$$0 = \text{completely self-incompatible}$$

Table 7. A schedule of procedures for assessment of the breeding system and their implications

Test		Treatment		Remarks
1 Control	Unbagged	Untreated	Free pollination	Evaluation of pollination rate under natural conditions
2 Spontaneous self-pollination	Bagged	Untreated	Pollen source: the same flower	Measuring the need for pollinators
3 Induced self-pollination	Bagged	Emasculated	Pollen source: different flower of the same plant	Indication of self-compatibility/incompatibility systems. Pollinators may induce self-pollination which does not occur spontaneously
4 Geitonogamy (artificial)	Bagged	Emasculated	Pollen source: another flower of the same plant	
5 Cross-pollination (artificial)	Bagged	Emasculated	Pollen source: another individual of genet	Indication (with item No. 4) of existence of compatibility systems.
6 Cross-pollination under natural conditions	Unbagged	Emasculated	Free pollination	Comparison of the results of 5 and 1 to 6, will determine if the seed set is limited by the pollinator
7 Control (for emasculation).	Unbagged / Bagged	Emasculated	Free pollination / Artificial pollination	Check for possible influences of emasculation
8 Agamospermy	Bagged	Emasculated	Without pollination	Evaluation of floral non-sexual reproduction rate. Excluding cases in which pollination promotes hormonal activity to produce seeds but without fertilization

Bagged = flower covered prior to flowering and throughout the flower's life-span, or in insect-free net houses.
Emasculated = excised anthers before dehiscence and keep the flower bagged throughout the floral life-span.

Table 8. Parameters used to measure reproductive success

Parameter	Procedure	Remarks
Pollen germination on the stigma	Examined (or fixed) several hours after natural or artificial pollination	Positive results may mask post-zygotic isolation mechanisms and late-acting incompatibility (93)
Growth of pollen tubes	Examined (or fixed) several hours after pollination	Comparison between number of pollen grains present on the stigma and number of pollination tubes may indicate the existence of incompatibility
Fruit set	Counting fruit number as a result of the treatment (% of the test flowers) at the end of the reproduction cycle	Seed set *per se* may be high but seed number/weight and/or viability may be low
Number or weight of seeds per fruit	Weighing and counting seeds per fruit	Number of seeds may indicate quantitative success but seed weight may indicate a qualitative one. Usually these two parameters are negatively correlated
Seed viability	Germination or tetrazolium test	The most reliable criterion for reproductive success

The rates of fruit set following artifical self-pollination have been ranked as follows:

Self-incompatible   0–3%   class 0
Slightly self-compatible 3–30%   class 1
Highly self-compatible   >30%   class 2

The ISI values should be considered cautiously since Kenrick (23) has shown that the level of ISI varies between individuals of the same species as well as among species. Seed set counting has also to include a test for seed viability. Furthermore, false estimation of the ISI may occur, since the absence of seeds may result from the sterility of ovules, although the pollen tubes are quite compatible (24).

3.2 Estimating the selfing rate

The frequency of the self-pollination should be estimated by comparison of the values of seeds from naturally pollinated flowers with those of hand self-pollination and hand cross-pollination. Writing P_x for the value from crossing, P_s for the value from selfing, and P_o for the open pollination value, the selfing rate, S, can be estimated as follows (25):

$$S = \frac{P_x - P_o}{P_x - P_s}.$$

This method makes it possible to discern the rate of selfing under natural pollination conditions. Especially in large inflorescences a single pollinator may carry out cross- and self-pollination if self-incompatibilities systems are not evolved. The selfing rate (S) can also be defined as $1 - t$, where t is the outcrossing rate (26). This value of t does not exclude the possible effects of geitonogamy if it is compared with the outcrossing of emasculated plants exposed to free pollination, unless all flowers on the genet are emasculated.

4. Estimation of the outcrossing level

Cruden (27) established that the relation between the number of the pollen grains and ovules (the P:O ratio) reflects the breeding system as follows:

Cleistogamy	P:O range	2.7 –	5.4
Obligate autogamy	P:O range	18.1 –	39.0
Facultative autogamy	P:O range	31.9 –	396.0
Facultative xenogamy	P:O range	244.7 –	2588.0
Obligate xenogamy	P:O range	2108.0 –	195 525.0

The outcrossing level of a given species shows a tight association among the breeding system, the floral size, the temporary separation of anther dehiscence and stigma receptivity, and the spatial relationship of the stigma and the anther (27). Although, on a large scale, the pollen:ovule ratio reflects the breeding system, each case has to be studied in relation to its specific pollination syndrome. Recently, there has been growing evidence (*Table 9*) which shows deviated P:O ratio from the standards suggested by Cruden.

Protocol 3. Estimation of pollen grain number per flower

Materials

- 70% ethanol
- detergent
- 0.5% methylene blue

Method

1. Squash a ripe anther (just before pollen exposure) into 0.9 ml ethanol + 3 drops of dye + 4 drops of detergent.
2. Transfer the squashed anther (carefully rinse the forceps and the blade in the droplet to prevent loss of material) into a calibrated tube and fill up to 1 ml with the same ethanol/detergent solution.
3. Shake well (or stir the suspension with a vortex mixer for 60–90 sec).
4. Transfer six separate samples of 1 μl each and count all pollen grains (this can be done with a haemacytometer).

5. To obtain the total number, calculate the average number of the pollen grains per anther based on the dilution factor, and multiply by their number in the flower.

Variations on this general procedure, especially in thick-walled anthers, may include:

- Softening of the anther tissue in 2 N KOH or 1 N HCl. The length of the softening period depends on the thickness of the theca wall and may even extend to 12 h at room temperature.
- Fixation in ethanol:glacial acetic acid 3:1 (especially for long storage before examination).
- Maceration in lactic acid:glycerol solution 3:1.
- Use of various other stains: aniline blue, basic fuchsin.

Table 9. Main sources of deviation of P:O ratios from the standards of Cruden (7)

Category	Causes and implications
Evolutionary history	Each genus seems to have a particular range of P:O for autogamous and for xenogamous species (28). The standards for evaluating breeding systems based on P:O has to be set on the level of the family (29)
Pollination mechanisms	Large pollen units and high ovule numbers (low P:O) are correlated with sophisticated exact pollination while low pollen unit and few ovuled (high P:O) related to wind on non-specialised pollination (33). Existence of 'feeding' (false) and 'pollinating' (true) anthers (34). In legumes, species with explosive mechanisms produce less pollen than congeners and other mechanisms (35)
Intraspecfic differentiation	Occurrence of autogamous and xenogamous populations (36) or individuals (37)
Intra-plant level	Co-existence of staminated and bisexual flowers (38). Flower position in the inflorescence (29). Changes in P:O during the flowering season (39)
Other factors	P:O ratios may be correlated with or result from: (a) resource allocation to sexual functions (30) (b) seed size, sex ratio, and pollen deposited on the stigma (31) (c) stigmatic pollen germination (32)

4.1 Ovule counting

Materials: Cotton blue–lactophenol (see Appendix A2).

Method: Crush the pistil or split it gently longitudinally with a blade on a glass slide in cotton blue–lactophenol and count the ovules under a dissecting microscope or hand lens. In many species the ovules need not be stained at all. In general, it is sufficient to isolate the ovary and to count the ovules.

Cruden (27) suggested an outcrossing index (OCI) which is expressed as the sum of the values for the three characteristics of flower size and behaviour, as follows:

- diameter of the flower (or of the inflorescence)
  up to 1 mm = 0
  1–2 mm = 1
  2–6 mm = 2
  >6 = 3
- temporal separation of anther dehiscence and stigma receptivity
  homogamy, protogyny = 0
  protandry = 1
- spatial positioning of the stigma and the anthers
  same level = 0
  spatially separated = 1

Note that by lumping homogamy and protogyny we ignore the role of protogyny in promoting outcrossing (39).

OCI values:
  0 = cleistogamy
  1 = obligate autogamy (certain weeds)
  2 = facultative autogamy (some outcrossing, local colonizers)
  3 = self-compatible, some demand for pollinators
  4 = partially self-compatible, outcrossing, demand for pollinators

The OCI value may be regarded only as a general indication of the breeding systems and should not be used as a proof. It is a useful field method, especially when many plant species are involved in the survey.

5. Pollen travel and gene flow

5.1 Estimation of pollen travel distance

The pollen travel distance is frequently used as a measure of gene flow, especially when detailed genetic analyses (e.g. of allozymes) are not available. With pollen flow measurement (or pollen mimic) actual gene flow via the pollen can only be estimated. The gene flow can be influenced by the incompatibility system involved, gamete competition (41), or optimal distance of crossing (42). As a result, actual gene flow may depart from pollen flow (ref. 3, p. 175).

Pollen may travel long distances, a phenomenon which is well-documented (43) but little information is available on intraspecific pollen travel between individuals. Pollen marking techniques provide information on travel distance (see *Table 10*). This knowledge is essential for the understanding of the genetic structure of the population.

Regardless of whether pollination mode is anemophilous or zoophilous, pollen travel has a leptokurtic distribution. This situation results from the clumped distribution of the plants, the population reward structure, and pollinator behaviour, as well as the aerodynamic properties of the pollen grains (ref. 3, pp. 167–73; ref. 44). Direct methods for estimating pollen travel distance involve physical, chemical, or genetical marking of pollen grains, while in the indirect method pollen analogues (powder dyes) are substituted for real pollen. In several rare cases (e.g. heterostyly), morphological differences in pollen size, form, colour, or exine microstructure are used to identify the pollen source. The distance and directionality of pollen travel are of crucial importance in evaluation of the pollinators' role in the rate of gene flow, and competition between plant species (e.g. by clogging of stigmas by improper pollen). The directionality and magnitude of pollen flow may influence the genetic subdivision of the population by its restrictions on their chances of panmictic mating.

In highly dispersed and/or rare plants, knowledge of the pollen travel distance (e.g. 'trapline' pollination routes) is especially important in any management or conservation scheme. The pollen removal:receipt ratio may indicate efficiency of both pollinator and pollination and the ratio of wasted pollen in relation to ovule number and success. The timing of pollen removal/ deposition is crucial, especially in dichogamous species, and may be a factor in the efficiency of sexual selection.

Assessment of compatible/incompatible pollen on the stigma is limited to heterostylous species, in which self/non-self pollen may be indicated if the pollen grains are polymorphic for colour or other traits. In these cases, the pollen tracing may be accompanied by manipulations of the population (natural or artificial), regarding the proportion of the morphs and the distances among them.

The exact location of pollen on a pollinator's body may indicate whether the pollen will reach its target or be wasted. In heterostylous species, exact locations of pollen deposit on the visitor's body in relation to the likelihood of contact of these parts with the proper stigma is a guide for distinguishing between a flower visitor and a true pollinator. One agent may visit several species and serve successfully as a pollinator of all of them by carrying the pollen on different parts of its body. Neotropical orchids and their euglossine pollinators are a good example (45).

Identification of the pollen source on the forager's body is sometimes a by-product of pollen staining or of the use of a fluorescent dye (but may also be achieved by pollen sampling and identification). The existence of dyed pollen on the pollinator or the stigma of other plants/species may be used in intra- and interspecific competition studies (between pollinators as well as between

Table 10. Methods for tracing and evaluation of pollen flow, their advantages and disadvantages

Method	Advantages	Disadvantages	Applications[a]
Pollen vital staining	Non-expensive, non-destructive. Individual pollen grains can be traced. Handy and useful in field experiments	Limited to small-scale experiments and to species in which the exposed pollen can be stained *in situ*	a, b, c, d, h
Powder dye as pollen mimic	Easy to handle, inexpensive, allows large samples. Fluorescent dyes are better detected after dispersal. Enables field manipulations involving pollinators or their dummies. The use of various colours simultaneously enables reciprocal trace of pollen flow and various manipulations	Possible low correlation between the powder movement and the real pollen transport. For gene flow studies, it is limited to a few species; needs calibration for each species	a, b, c, d, g, h
Radioactive labelling of pollen grains	Pollen is definitely marked and not disturbed in their natural dispersal, individual grains could be traced and scored. Gives accurate description of the real pollen flow	Expensive method, possibly environmentally hazardous, complicated technically, practically almost abandoned. Needs large samples of stigmas as a function of the distance from the marked plant	b, c
Pollen surface labelling followed by neutron activation analysis	Non-destructive method, very precise and very small amounts of pollen can be detected. Samples can be analysed at any given time	Complicated technically and expensive practically, seldom used. It is assumed that topical application of the labelling material does not interfere with the aerodynamics of the pollen	b, c
Genetic markers	Very accurate; allow analysis of large samples	Limited to easily recognized dominant genes or allozymes. (Then needs electrophoretic analysis)	b, c, h

Polymorphism in pollen colour	Pollen can be easily detected on the stigma; the system enables manipulation of the pollen source	Limited to very few species	a, c, d, f, h
Heteromorphic pollen	Possibility of differentiation between the pollen morphs	Limited mainly to heterostylous species	a, b, c, e, f
Progeny analysis	Accurate and controlled system which enables experimental manipulation	Dominant marker gene; needs controlled breeding and experimental growing of the plants; generally not applicable for natural populations	b
Pollen species-specific morphology	Used to detect and quantify inter-specific pollen flow	Closely related species often have similar pollen grains	a, b, c, g, h, i

[a] Corresponds to the list on p. 44.

plants). In rare species or infrequently visited ones, pollinators can be indirectly identified, especially by the wide use of a fluorescent dye in the application of an intensive capture programme (e.g. by the random sweeping of insects or by using mist nets for birds).

5.1.1 Pollen marking and tracing

Pollen marking and tracing have been used extensively for various purposes under field conditions, and have been accompanied by other observations or manipulations, some of which are very expensive (e.g. neutron activation of pollen grains (44, 46) or 'labelling' of pollen with radioactive elements (47, 48)) which are impractical for extensive use. Other methods (e.g. the use of pollen colour and shape or size polymorphism) are limited to a few specific taxa or require cultivation and the expression of easily recognized dominant genes (e.g. the use of genetic markers). From a practical point of view, pollen staining and the use of dyes are the easiest and cheapest methods for extensive use (44, 45, 49). *Table 10* summarizes the use of different marking methods and their applications.

Pollen grains are marked and traced for the following purposes:

(a) to assess the pollen carry-over

(b) to estimate the genetic neighbourhood size

(c) to determine pollen travel distance and directionality

(d) to estimate the pollen removal:pollen receipt ratio

(e) to assess the compatible:incompatible pollen ratio on the stigma

(f) to assess the self:non-self pollen ratio on the stigma

(g) to localize pollen on the pollinator's body

(h) to determine effectiveness of pollen removal by the pollinator

(i) to identify previously unnamed flower visitors and the sharing of pollinators

Studies of pollen carry-over measure the number/proportion of pollen grains deposited on a virgin and receptive stigmas after a known number of subsequent pollinator visits. The pollen carry-over value may determine the proportion of geitonogamy in relation to cross-pollination. It is also essential to consider the number of flowers visited per plant, the proportion of co-occurring receptive stigmas, and the compatibility systems of the plant (3). Manipulations are frequently applied to force the pollinator to follow a given route of test flowers prepared in advance, and to eliminate the non-controlled ones. The estimation of the genetic neighbourhood size is based mainly on genetic markers (6), which demands a long-term procedure, an intimate knowledge of the plant's genetic make-up, and easily recognizable genotypes. The rare cases of variations in pollen colour (50, 51) or form dimorphism (52) are also applied.

Protocol 4. The use of fluorescent dye to mimic pollen

Materials

- fluorescent powdered dye (e.g. Black Ray®, Day Glo®, or Radiant Color® or ordinary dye powder (methylene blue, carmine acetate, neutral red, Evans blue, Bismarck brown)
- toothpick or fine paint-brush

Method

1. Apply (with a toothpick or fine brush) a small amount of the powdered dye to the pollen exposure surface of the freshly dehisced anther. The exact quantity of powder should be such that no other part of the flower be contaminated, especially those which may be touched by flower visitors during their stay.
2. Depending upon the purpose of the inquiry, different colours may be used simultaneously in different flowers, sexes, individuals, species, etc.
3. Harvest the target stigmas (in planned manipulations of known flowers or plants) or stigmas of other flowers around the dye source, and store each stigma separately in melted gelatine (or another colourless preserving medium, see Chapter 3, *Protocols 7–9*). Note the sources of the stigma (species, flower position on the plant, distance and direction from the dye source).
4. Examine the excised stigma under an epi-fluorescent (if fluorescent dye was applied) microscope or with hand lens (for regular dye particles) for the presence of the dye particles.

While working with wild non-manipulated plants, a survey for the detection of dye fluorescent dispersal is done by using a portable UV lamp at dusk. The efficient discovery distance depends on the lamp's intensity and is normally between 30–100 cm. Fluorescent stigmas can be fairly easily screened in large masses. Furthermore, pollinators can be detected, especially if they are sleeping in the flowers. By this method, the amount of harvested stigmas can be minimized to the researcher's purpose.

Hessig (53) introduced another method; the dyes were placed on stigmas rather than anthers. The stigmas are more traceable than anthers and deposited dyes on them will be attached to the visitors. The advantage of this method is the immediate screening between foragers that may not touch the stigmas and real pollinators.

The dye particles may differ from pollen grains in their specific weight, size, and form, and thus in their adhesive capability and/or buoyancy (54, 55). These differences should be kept in mind when analysing the results. This method is especially efficient in studies of sticky pollen to which the powdered dye will easily adhere (56).

While using dyes as analogues to pollen there is a need to examine experimentally that the dye will not interfere with post-pollination success by clogging stigmas. Application of dye prior to hand-pollination, either to anthers in a staminated-phase flower or directly on the stigma, in comparison to only hand-pollination as a control, will uncover such an influence (57).

Protocol 5. Staining of pollen grains *in situ*

Materials

- vital dye liquid (methylene blue, 0.5%; neutral red, 0.5%; brilliant green, 1%; Bismarck brown, 1%; orange G, 10%; rhodamine, 0.2%; tryphan red, 2%)

- a toothpick or fine paint-brush

Method

1. Apply a small amount of the liquid dye with a toothpick or fine brush on to the pollen exposure surface of the freshly dehisced anther. The dye quantity depends on the theca size, spatial position, pollen amount, etc. Be careful not to stain any other parts of the flowers, especially those which the visitor usually touches during its stay on the flower.

2. Check carefully that the dye does not injure the pollen. Examine the rate of bursting and pollen stickiness (see Chapter 7, *Protocol 3*) after the dye has dried. Catch the potential pollinator and examine it for the presence of dyed pollen. All these controls are essential prior to any examination of pollen travel distance; if the dye changes the pollen dimensions or dispersability, this method should be discarded (61).

3. Use different colours simultaneously on different flowers, sexes, individuals, and species, depending on the purpose of the inquiry.

4. In orchids and asclepiads the whole pollinium can be stained and traced as a complete unit (49).

5. Trace the pollen dispersal and directionality by harvesting the stigmas around the marked flower. Remove every stigma and store each one separately (in FAA or 70% ethanol) until examination. Mark the source of the stigma clearly (species, flower position on the plant, distance from the dye source, time since the marking, etc.).

6. After squashing, examine the excised stigmas through a microscope and count the number of stained and non-stained pollen grains.

7. Calculate the ratio of stained:non-stained pollen grains as a function of the distance from the dyed pollen source and the time since marking.

5.1.2 Pollen carry-over

Pollen carry-over refers to transfer from a male-phase donor flower to a series of female-phase flowers (ref. 3, p. 160, and ref. 54). Pollen carry-over measurement is an indirect method for investigating the mating system of plants to replace more complicated genetic analyses of parents and offspring (44) or electrophoretic allozymes analysis (see ref. 6 for review) which may encompass only a fraction of the potential pollen parents. Pollen carry-over studies relate the actual pollen load (of a known source) to the rate of ovule fertilization and to the resultant seed production (58).

The rate of pollen carry-over and the pollen deposition decline curves along a single foraging bout may indicate the pollination efficiency of a particular pollinator and the number of visits needed to ensure full seed production. Carry-over (C_x) is characterized by the number of pollen or dye grains (Y_x) divided by the number deposited on the first flower in sequence, $Y_1/[C_x = (Y_x/Y_1)]$ (for successively visited flowers $x = 2, 3, 4, 5, 6$), assuming that the decay in pollen deposition is exponential (59) according to known results (see ref. 60, for example). Exponential decay results if a pollinator deposits a constant percentage (K) of the amount carried on its body in a given visit. This assumption seems valid only if in between two successive visits there is not any grooming behaviour which may change the pollen distribution on the pollinator's body and hence also the chances to load a certain proportion of pollen on the next conspecific stigma. Another critical assumption is that all the flowers or inflorescences on a given plant have the same chance to be visited by a pollinator and that the pollinator handle all the flowers in the same way regardless of the plant or flower size or the available reward.

Protocol 6. Pollen carry-over estimation

Materials

- test plants in enclosures or a greenhouse
- tags for marking individual flowers
- a live and active pollinator

Methods

1. Transfer the experimental plants to an insect-free enclosure or greenhouse prior to anthesis. If some flowers are present, remove them.
2. Emasculate the flower while in bud to ensure virgin stigmas without any pollen (see pp. 33–34).
3. Introduce a pollinator (bee, bird) into an enclosure with donor plants, let it acclimatize in the enclosure before running the experiment, and then let it forage on the plants.

Protocol 6. *Continued*

4. Transfer the pollinator to a second enclosure (greenhouse) into which the emasculated flowers have been placed.
5. Note each visited flower and mark its position in the foraging bout sequence.
6. Excise each visited stigma according to the order of visit and count the pollen grain load (see *Protocol 3*).
7. Run the same procedure with more individuals or with other pollinators.
8. Measure the following parameters for each foraging bout: number of flowers visited, number of pollen grains per flower in relation to its position in the visiting sequence, the average distance between flowers in each bout. Use this data to produce the pollen carry-over decay curve and to estimate the distance of pollen transfer.

Variation on this general procedure may include: use of a dead animal [insect (60) or bird (61)]. The animal is artificially loaded with pollen in a manner as close as possible to its natural position on the flower and then is transferred sequentially into virgin flowers in the same way. The stigmas are then harvested and the pollen load counted. If the seeds are later collected for examination of offspring fitness, the stigmas are checked intact by a portable microscope (×50) for evaluation of the pollen amount and then re-covered again with a bag for seed production.

A common method for simulation of pollen travel is to use fluorescent dye powder as a pollen mimic (*Protocol 4*). Waser (54) realized that different pollinators may transfer dye differently as compared with the transfer of real pollen, but on average pollen and dye transfer are similar (54, 55, 62).

In some experiments, birds were trained to manipulate the flower and to use it as a nectar source prior to the running of the experiment (63, 64). While working with insects (*Bombus* or *Psithyrus*), Thompson and Plowright (65) found that the animals need an acclimatization period and that a suitable source of nectar for their own consumption must be provided. If the same animal is used for several runs it must be cleaned of all the dye powder or pollen prior to examination to prevent contamination (63).

Waddington (60) dyed the pollen *in situ* (with basic fuchsin in 95% EtOH) and later (45 min) introduced a dead bee to be loaded with pollen and then to a sequential series of stigmas. By this simple and rapid method one can overcome the need to isolate stigmas or to train the pollinator to learn to manipulate the particular test plants (65) and the results are instant. Of course the results should be taken with reservation since the behaviour of a live bee could disperse pollen in a manner which deviates from the simulation of dead ones.

Experiments on pollen carry-over may be carried out on natural unmanipulated plants in relation to natural behaviour and movements of the pollinators (see ref. 66, for example) in order to predict the pollination efficiency (as expressed by viable seeds; see Chapter 6, *Table 4*) in relation to the pollen source. Pollen carry-over studies supply information on pollen travel and thus on the genetic make-up of the population. It should be noted, however, that the real importance of carry-over can only be interpreted with reference to the number of receiving flowers on the donor plant, the number of flowers on subsequent recipient individuals, and the compatibility system of the plant (ref. 3, p. 163). The reward structure will dictate the flight distance and thus, also the rate of geitonogamy in self-compatible species (67).

Under natural conditions, pollen reaching the stigma may frequently be of various genetic origins leading to multipaternity of seed in the same fruit (68). Any pollen carry-over analysis should consider this phenomenon in the interpretation of the genetic consequences of pollen transfer.

5.1.3 Optimal outcrossing distance

Reproductive success in many plant species is influenced by the degree of inbreeding and outbreeding. A high level of inbreeding may result in a reduced number and quality of offspring, mainly in predominantly out-breeders (ref. 3, pp. 359–63; ref. 25, p. 74).

Outbreeding depression may occur in crosses between separate populations of the same species (69) or even within a single population (70). Shields (71) explains this phenomenon as a by-product of a highly local adaptation while Waser and Price (72) explain it as an intragenomic co-adaptation in the absence of spatially varying selection.

Waser and Price (72) noted that if both inbreeding and outbreeding depression occur within a population, there should be an intermediate degree of outbreeding at which its overall deleterious effect is minimized (70, 71). This in turn will correspond, on the average, to an optimal outcrossing distance, given that genetic similarity declines with distance (72).

Protocol 7. Estimation of the optimal outcrossing distance (70)

Materials

- pollination bags
- tags for marking of individual flowers

Method

1. Bag the test plants at the bud stage to exclude any pollen on the stigma.
2. Hand-pollinate receptive virgin stigmas of the test flower with pollen brought from donor plants of the same species growing at various known distances (e.g. 0, 10, 30, 100 m) from the recipient plants.

Protocol 7. *Continued*

3. Bag the hand-pollinated flowers immediately after pollen deposition on the stigmas to prevent any pollen addition or contamination.
4. Clearly mark each flower for each treatment.
5. Harvest the seed set after ripening. Count and weigh the seeds of each treated flower according to the various distances from the pollen source.
6. Sow the seeds preferably in natural habitat or in experimental plots or a greenhouse and compare the relative average performance (e.g. seedling growth rate, plant size, flower number) of seeds according to distance of the donor pollen source.
7. Use the data of the seed number, weight, and performance to estimate optimal outcrossing rate based on the relative fitness of the offspring.

To check optimal outcrossing distance choose a species with a well-known flower life-span to ensure the use of fresh pollen and the adequate receptive stigmas (see Chapter 3, Section 6). Knowledge of the breeding system (Section 1 above) and rate of self-incompatibility (Section 2.2) is essential for any further interpretation of the results.

Standardize the pollen application techniques on the stigmas to ensure deposition of similar amounts of pollen in each treatment and the time which elapses between pollen removal from the donor source and the pollen deposit on the recipient (76).

Avoid working under extreme weather conditions (too wet or too hot) which may influence pollen performance (73).

In all the replicates leave several unpollinated flowers to serve as a control for possible failure of the exclusion bags. If control flowers on any plant set more seed than the average for pollinated flowers from the replicate as a whole, exclude this plant from the analysis (70). A species whose seeds show dormancy is less suitable for outcrossing distance study because seed performance analysis becomes more complicated and expands over a long time period. Waser and Price (72) also include the examination of offspring (of various outcrossing distances) fitness as revealed by the performance of the seedlings in the maternal environment and in the herbaceous vegetation typical of the test species habitat.

6. Mating patterns and sexual selection

6.1 Sexual selection, male and female fitness

For many years ecologists have considered the reproductive success of plants entirely in terms of seed production, which only reflects the maternal investment. In reality every seed genome is half maternal and half paternal and on the average, in population terms, male and female contributions to the next generation are equal. Hence, an evolutionary interpretation of reproductive characters must consider their effects on both male *and* female performance (39).

Stephenson and Bertin (74) define sexual selection as the differential reproductive success of individuals of the same sex and species that survive to reproductive age and are physically capable of reproduction. According to this definition, which eliminates any considerations of survival, intrasexual (usually male) competition (*Figure 1*) and intersexual (usually female) choice

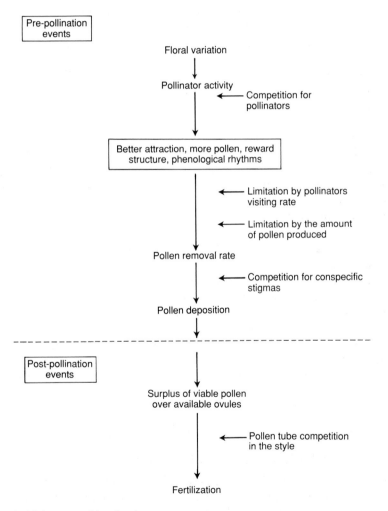

Figure 1. Male competition in plants.

(*Figure 2*) are the primary causes of sexual selection. Two main mechanisms may be involved:

- more pollen from one donor is received than from other donors in the same population
- non-random fertilization or abortion occurs following pollen reception

In a broad sense, if resources are limited, the selection on the male and female reproductive function will differ considerably. For example, if there is a surplus of male gametes available but resources are limited, the female can enhance its fitness by its choice of a male gamete or zygote, presumably with better genetic constituents (75). Selective fruit abortion can improve the quality of seed production (76). If resources are not limited and there is access to female gametes, selection will favour adaptations promoting pollen removal such as larger advertisement (inflorescences as well as flower size; but see ref. 77) and temporal spread of pollen presentation (74).

The male function hypothesis states that large inflorescences achieve fitness not only as females by setting more fruits but also as males by donating more pollen for fertilization (78, 79). In line with this hypothesis, success in competition for pollen delivery, which means fitness gained through the male function, increases with the number of flowers. Fitness via female function is reduced with increasing flower number because of resource limitation on seed set. Thus, the selective advantage of large inflorescences is through the male function (ref. 74; but see ref. 77) by donating more pollen. This hypothesis predicts that pollen donation increases more rapidly than seed production as flower number increases.

6.1.1 Approaches to the study of male and female fitness

The first step in checking male and female fitness of hermaphroditic flowers is to test whether seed set is pollen- or resource-limited. Controlled hand-pollination (80, 81) or careful observation of natural pollen deposition rate (82, 83, 84) in relation to seed-set rate, may reveal if there is an adequate amount of pollen.

If seed production is not pollen-limited the research is focused on pollen removal and deposition in relation to phenology, flower number and arrangement, proximate attractants, and floral rewards. Post pollination competition includes the *in vitro* and *in vivo* male gametophytic performance (competition at the style) (see ref. 74 for review). Campbell (85, 77) noted that success in pollen removal may not necessarily be directly proportional to success in delivery to stigmas. Simultaneous pollen removal and pollen donation in relation to the inflorescence/flower size must therefore be checked.

If seed production is resource-limited, female choice is regarded as the main factor in selective seed abortion (*Figure 2* and *Table 11*). Thus the research is focused on the survivorship of fertilized ovules (at different

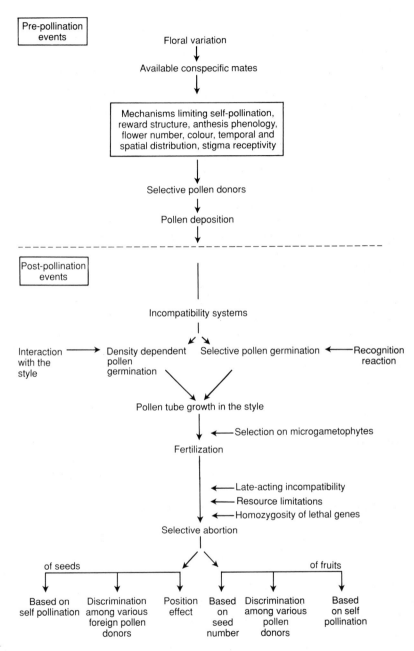

Figure 2. Female choice in plants.

Table 11. Main components of male and female fitness and their measurement

Component	Method	Remarks
Pollen removal rate	Monitoring the rate of pollinia removal in Orchidaceae and Asclepiadaceae (86, 90). Estimation of removal rate of marked pollen grains or their analogues	Success in pollen removal may not be directly proportional to success in pollen delivery (92). Studied mainly in relation to inflorescence size, visitor frequency, and nectar production
Pollen delivery and deposition	Carry-over studies. Monitoring the rate of pollen delivery, marked pollen, or pollen analogues. Examination of the pollen on the forager	Pollen carried by foragers can depart considerably from pollen delivery
Rate of pollen germination	Counting the pollen germination rate on stigmas at various times	Population density effect should not be neglected
Pollen germination and tube growth	Study in relation to incompatibility systems	
Pollen tube competition	Monitoring pollen tubes in the style in relation to the ovule number	
Seed and fruit abortion	Counting the survivorship and fertility of ovules, developing embryos, seed, and fruits abortion number along various stages of the fruit development	The data are used to calculate the Relative Reproductive Success (RSS) (92). $$RRS = \frac{fruit}{flowers} \times \frac{seeds}{ovules}$$ Infra-fruit competition as a result of resource limitation may influence the final seed number per fruit or the seed spacing in the fruit. A non-limited pollination situation has to be maintained to eliminate ovule competition for males

development stages), seeds and fruits in relation to life history, breeding systems, and reproductive success (86); paternity patterns (see ref. 39 for review), seed number per fruit ('the selective abortion hypothesis' (76, 87); ovules position effect (88) and self- and cross-pollination (89)). Schemske and Paulter (90) realized that four components of fitness should be studied simultaneously: seed production, seed weight, seed germination, and the seedling performance.

Male fitness and female fecundity have also been studied in relation to flower colour polymorphism in a genetically known experimental population (ref. 91 and earlier works), and under natural conditions by reciprocal careful monitoring of the removal and deposition of different dye particles (85). These two studies exemplify the need for a comprehensive study which considers the male and female aspects simultaneously, an approach that minimizes non-tested assumptions.

Recently, Stanton and Preston (92) realized four possible cases of seed production (resource and/or pollen deposition limitation) in relation to pollen export (visit or pollen-pool amount limitations):

(a) resource-limited and visit-limited;

(b) resource-limited and pollen-pool-limited;

(c) pollen-pool-limited and visit-limited; and

(d) pollen-limited and pollen-pool-limited.

Each of these four cases has different implications on seed number and quality, rate of pollen production, and pollen export. Thus, it seems that the dichotomy into pollination-limited or resource limited seed production is an over-simplification and should be studied under visit-limited as well as under pollen-pool-limited environments stressing the dynamics of pollen export and paternity (91).

Another precaution comes from the study of Wiens (86) who noted that pollen and resource limitations are not the exclusive causal factors for brood size (mean number of seeds matured per ovary) reduction, since the brood size is correlated with different modes of dispersal, strategies of ovule packaging, and life forms.

References

1. Wyatt, R. (1983). In *Pollination Biology* (ed. L. Real), p. 51. Academic Press, Harcourt Brace Jovanovich, Orlando, Florida.
2. Frankel, R. and Galun, E. (1977). *Pollination Mechanisms, Reproduction and Plant Breeding*. Springer-Verlag, Berlin.
3. Richards, A. J. (1986). *Breeding Systems*, Allen & Unwin, Winchester, MA.
4. Barrett, S. C. H. and Eckert, C. G. (1990). *Isr. J. Bot.*, **39**, 5.
5. Ornduff, R. (1969). *Taxon*, **18**, 121.
6. Brown, A. H. D. (1989). In: *Population, Genetics, Breeding, and Germplasm Resources* (ed. A. H. D. Brown, M. T. Clegg, A. L. Kahler, and B. S. Weir), p. 17. Sinauer, Sunderland.
7. Cruden, R. W. (1977). *Evolution*, **31**, 32.
8. Bawa, K. S. and Beach, J. H. (1981). *Ann. Missouri Bot. Gard.*, **68**, 254.
9. Lloyd, D. G. (1980). *Bot. Monog.*, **15**, 67.
10. Faegri, K. and L. van der Pijl. (1979). *The Principles of Pollination Ecology* (3rd edn). Pergamon Press, Oxford.

11. Webb, C. J. and Lloyd, D. G. (1986). *NZ J. Bot.*, **24**, 163.
12. Nettancourt, D. de (1977). *Incompatability in Angiosperms.* Springer-Verlag, Berlin.
13. Nettancourt, D. de (1984). In: *Encyclopedia of Plant Physiology* (ed. H. F. Linskens & J. Heslop-Harrison). New Series 17:624. Springer-Verlag, Berlin.
14. Charlesworth, D. (1987). *Persp. Biol. Med.*, **30**, 263.
15. Shivanna, K. R. and Johri, B. M. (1985). *The Angiosperm Pollen-Structure and Function.* Wiley Eastern, New Delhi.
16. Ganders, F. R. (1979). *NZ J. Bot..*, **17**, 607.
17. Barrett, S. C. H. (1990). *TREE*, **5**, 144.
18. Cruden, R. W. and Lyon, D. L. (1989). In *The Evolutionary Ecology of Plants* (ed. J. H. Bork and Y. B. Linhart), p. 171. Westview Press, Boulder, Colorado.
19. Bell, G., Giraldeau, L. A. and Weary, D. (1984). *Oecologia (Berl.)*, **64**, 287.
20. Young, T. P. (1982). *Am. J. Bot.*, **69**, 1326.
21. Guo, Y-H. and Cook, C. D. K. (1990). *Aquat. Bot.*, **38**, 283.
22. Zapata, T. R. and Arroyo, M. T. K. (1978). *Biotropica*, **40**, 221.
23. Kenrick, J. (1986). In *Pollination 86* (ed. E. G. Williams, R. B. Knox, and D. Irvine), p. 116. Plant Cell Biology Research Centre, School of Botany, University of Melbourne, Parkville.
24. Gaude, T. and Dumas, C. (1987). *Int. Rev. Cytol.*, **107**, 333.
25. Charlesworth, D. and Charlesworth, B. (1987). *Annu. Rev. Ecol. Syst.*, **18**, 237.
26. Karron, J. D. (1987). *Ecol. Evol.*, **1**, 47.
27. Cruden, R. W. (1977). *Evolution*, **31**, 32.
28. Vasek, F. C. and V. Weng (1988). *Syst. Bot.*, **13**, 336.
29. Preston, R. E. (1986). *Am. J. Bot.*, **73**, 1732.
30. Charnov, E. L. (1982). *The Theory of Sex Allocation.* Princeton University Press, Princeton, NJ.
31. Shanker, U. and Ganeshaia, N. K. (1984). *Curr. Sci.* **53**, 731.
32. Plitman, U. and Levin, D. A. (1990). *Pl. Syst. Evol.*, **170**, 205.
33. Kress, W. J. (1981). *Syst. Bot.*, **6**, 101.
34. Vogel, S. (1978) In *The Pollination of Flowers by Insects* (ed. A. J. Richards, p. 89. Academic Press, London.
35. Small, E. (1988). *Pl. Syst. Evol.*, **160**, 195.
36. Schoen, D. J. (1977). *Syst. Bot.*, **2**, 280.
37. Wyatt, R. (1984) *Syst. Bot.*, **9**, 432.
38. Lindsey, A. H. 1982 *Syst. Bot.*, **7**, 1.
39. Bertin, R. I. (1989). In *Plant–Animal Interaction* (ed. W. G. Abrahamson), p. 30. McGraw-Hill, New York.
40. Lord, E. M. (1980). *Am. J. Bot.*, **67**, 1430.
41. Mulcahy, G. B. and Mulcahy, D. L. (1975). *Theor. Appl. Gen.*, **46**, 277.
42. Waser, N. M. and Price, M. V. (1983). In *Handbook of Experimental Pollination Biology* (ed. G. E. Jones and R. J. Little), p. 341. Van Nostrand Reinhold, New York.
43. Faegri, K. and Iversen, J. (1989). In *Textbook of Pollen Analysis* 4th edn (ed. K. Faegri, P. E. Kaland, and K. Krzyrinski). John Wiley, Chichester and New York.
44. Handel, S. N. (1983). In *Pollination Biology* (ed. L. Real), p. 163. Academic Press, Orlando, Florida.

45. Williams, N. H. (1982). In *Orchid Biology Reviews and Perspectives II*. (ed. J. Arditti), p. 119. Cornell University Press, Ithaca, NY.
46. Gandreau, M. and Hardin, J. H. (1974). *Brittonia*, **26**, 316.
47. Turpin, R. A. and Schlising, R. A. (1971). *Radiat. Bot.*, **11**, 75.
48. Reinke, C. D. and Bloom, W. L. (1979). *Syst. Bot.*, **4**, 223.
49. Peakall, R. (1989). *Oecologia (Berl.)* **79**, 41.
50. Thompson, J. D. (1986). *J. Ecol.*, **74**, 329.
51. Thompson, J. D. and Thompson, B. A. (1989). *Evolution*, **43**, 657.
52. Barrett, S. C. H. and Wolfe, L. M. (1986). In *Biotechnology and Ecology of Pollen* (ed. D. L. Mulcahy and O. Ottaviano), p. 439. Springer-Verlag, New York.
53. Hessig, A. B. (1988). *Am. J. Bot.*, **75**, 1324.
54. Waser, N. M. (1988). *Funct. Ecol.*, **2**, 41.
55. Thompson, J. D., Price, M. V., Waser, N. M., and Stratton, A. (1986). *Oecologia (Berl.)*, **69**, 561.
56. Linhart, Y. B. (1973). *Am. Nat.*, **107**, 111.
57. Campbell, D. R. (1991). *Am. J. Bot.*, **75**, 1324.
58. Waser, N. M. and Price, M. V. (1991). *Ecology*, **72**, 171.
59. Campbell, D. R. (1985). *Evolution*, **39**, 418.
60. Waddington, K. D. (1981). *Oikos*, **37**, 153.
61. Pyke, G. H. (1981). *Anim. Behav.*, **29**, 889.
62. Campbell, D. R. and Waser, N. M. (1989). *Evolution*, **43**, 1444.
63. Price, M. V. and Waser, N. N. (1982). *Oecologia (Berl.)*, **54**, 353.
64. Waser, N. M. and Price, M. V. (1984). *Oecologia (Berl.)*, **62**, 262.
65. Thompson, J. D. and Plowright, R. C. (1980). *Oecologia (Berl.)*, **46**, 68.
66. Craig, J. L. (1989). *Oecologia (Berl.)*, **81**, 1.
67. Heinrich, B. and Raven, P. H. (1972). *Science (Wash.)*, **176**, 595.
68. Bertin, R. I. (1987) In *Plant Reproductive Strategies* (ed. J. Lovett-Doust and L. Lovett-Doust), p. 30. Oxford University Press.
69. Sobrevila, C. (1988), *Am. J. Bot.*, **75**, 701.
70. Waser, N. M. and Price, M. V., (1983). In *Handbook of Experimental Pollination Biology* (ed. G. E. Jones and R. J. Little), p. 341. Van Nostrand Reinhold, New York.
71. Shields, W. M. (1983). *Philopatry, Inbreeding and Evolution of Sex*, State University New York Press, Albany.
72. Waser, N. M. and Price, M. V. (1989). *Evolution*, **43**, 1097.
73. Corbet, S. A. (1990). *Isr. J. Bot.*, **39**, 13.
74. Stephenson, A. G. and Bertin, R. I. (1983). In *Pollination Biology* (ed. L. Real), p. 110. Academic Press, Orlando, Florida.
75. Willson, M. F. and Burley, N. (1983). *Mate Choice in Plants*. Princeton Biol. Monog. 19. Princeton University Press, Princton, NJ.
76. Stephenson, A. G. and Winsor, J. A. (1986). *Evolution*, **40**, 453.
77. Campbell, D. L. (1989). *Am. J. Bot.*, **76**, 730.
78. Willson, M. F. (1979). *Am. Nat.*, **113**, 777.
79. Sutherland, S. (1986). *Ecology*, **67**, 991.
80. Schemske, D. W. (1980). *Evolution*, **34**, 489.
81. Bierzychudek, P. (1981). *Am. Nat.*, **117**, 838.
82. Mulcahy, D. L., Curtis, P. S., and Snow, A. A. (1983). In *Handbook of*

Experimental Biology (ed. C. E. Jones and R. J. Little), p. 330, Van Nostrand Reinhold, New York.
83. Snow, A. A. (1986). *Am. J. Bot.*, **73**, 139.
84. Galen, C., Zimmer, K. A., and Newport, M. E. (1987). *Evolution*, **41**, 599.
85. Campbell, D. R., (1989). *Evolution*, **43**, 318.
86. Wiens, D., (1984). *Oecologia (Berl.)*, **64**, 47.
87. Lee, T. D. and Bazzaz, F. A. (1982). *Ecology*, **36**, 1974.
88. Horovitz, A., Meiri, L., and Beiles, A., (1976). *Bot. Gaz.*, **137**, 250.
89. Crowe, L. R. (1971). *Heredity*, **27**, 118.
90. Schemske, D. W. and Paulter, L. P. (1984). *Oecologia (Berl.)*, **62**, 31.
91. Stanton, M. L., Snow, A. A., Handel, S. N., and Bereczky, J. (1989). *Evolution*, **33**, 335.
92. Stanton, M. L. and Preston, R. E. (1988). *Am. J. Bot.*, **75**, 540.
93. Seavey, S. R. and Bawa, S. K. (1986). *Bot. Rev.*, **52**, 195.

3

Pollen and stigma biology

1. Introduction

The process of pollination begins with the exposure and shedding of ripe pollen which carries the male gametes or their progenitor. The pollen grain is exposed to a hostile environment (such as dry conditions) and has to reach the proper receptive stigma while it is vital.

Any failure of the pollen grain to germinate and later to fertilize an ovule means an unsuccessful result of a pollination event. Regardless of the pollen size, form, or mode of dispersal, the key factor is pollen viability, which may be estimated by various methods (see Section 2; *Table 1*). Evaluation of pollen viability is the first step in understanding the chances of a given pollen grain to germinate on the stigma as a crucial stage toward fertilization (*Table 2*). Pollen dose on the stigma (Section 7.1), although of the proper species and genotype, provides no information about its future fate.

The pollen–stigma relationship depends on pollen viability, stigma receptivity, and genetic interaction of both partners as dictated by the incompatibility system (if any).

Even a successful event of pollen germination does not ensure later success, due to various factors such as pollen competition (5), interactions with the style tissues (3, 6), late incompatibility processes, or post-zygotic abortion (7).

For most field pollination studies the main components to be regularly examined are pollen viability, stigma receptivity, pollen-tube growth in the style, and the breeding system. Stigma and style histochemistries elucidate important facets of pollen–stigma interactions (1) but are less applicable for most of the pollination projects.

The pollen grain has a wall which is not enveloped in cuticle but is surrounded by a complex exine. Two types of pollen grains are recognized in higher plants (*Table 3*):

(a) Bicellular type: In gymnosperms and in two-thirds of the flowering plant families. The pollen grain comprises two cells at maturity; that is, a vegetative cell which is related to the tube growth and metabolism, and a

Table 1. Methods for evaluating pollen quality *in vitro*

Method	Advantages	Disadvantages
Test of germinability *in vitro*	A real test of pollen viability; easy to carry out repetitively with large samples; easy to control the medium content; a retainable predictive test for pollen capability to germinate	Germination score may not necessarily correspond with the fertility capacity. The results are highly dependent on the medium content used. The test ignores the female partner's control of pollen germination *in vivo*
Test of stainability of the pollen cellular content	Easy to handle, cheap, permits large samples; useful for identifying high rates of pollen abortion (e.g. after hybridization)	May show low correlation with the real germinability *in vivo*
Tetrazolium test (TTC)	Rapid and fairly correlated with pollen viability, suitable for large samples	Pollen of different species are not equally sensitive to TTC tests. Desiccated pollen grains may give false-positive results
Fluorochromatic reaction (FCR)	Shows high correlation with the potential germinability (if mature pollen is used); reflects a real situation *in vivo*	Close correlation with germinability can only be expected when the optimum medium for the given species has been identified. Immature pollen may give false results. Pollen from male sterile plants may show positive results
Peroxidase reaction	Easy to carry out, gives a quantitative estimation. Suitable and quick for large samples	Unfresh pollen may give lower values of viability
Alexander's procedure to differentiate between aborted and non-aborted pollen	Clear distinction between fertile and aborted pollen grains	Somewhat complicated procedure which needs fine calibration for each specific case; pH-sensitive procedure

Sources: refs 12, 20, 25.

generative cell which divides within the pollen tube to produce two sperm cells.

(b) Tricellular type: In one-third of the angiosperms (e.g. Asteraceae, Poaceae, Brassicaceae, and Apiaceae). The generative cell divides precociously, so that all three cells are present within the pollen grain at maturity. The sperm cells are sited wholly within the cytoplasm of the vegetative cells, separated not by walls but by their own plasma membrane and that of the host cell.

Table 2. Pollen–pistil interaction (dry-type stigma)[a]

Stage	Controlling factors
Pollen arrival ↓	Pollination agents (biotic and abiotic). Weather conditions
Contact ↓	Surface patterning both of pollen as well as of the stigmatic papillae
Attachment and adhesion ↓	Chemical composition of the stigmatic and the pollen surfaces. Wind speed, surface tension, electrostatic and electrodynamic forces
Hydration ↓	Water supply from the stigma. Pollen apertural mechanisms (exine type and stigma-scaling devices)
Acceptance ↓	Recognition factors in stigma papillae
Germination ↓	Initial rehydration → aperture opening → full hydration
Tube penetration ↓ Tube growth ↓	Secretion in the transmitting tissues of stylar canal. Incompatibility systems
Fertilization	Ovule and pollen genetic constituents

[a] In wet stigmas contact and adhesion are indiscernible due to the sticky exudate. Sources: refs 1–4.

Mature pollen is dehydrated, with a low content of water, rehydration is needed to ensure the pollen tube emergence, and it is gained from the stigma (6, 13).

The stigmas and style are glandular organs, and their metabolism is tuned to the temporal processes of flowering and pollination. The stigma contains the receptive cells which recognize the pollen grain, and provides the germination substrate.

Stigmas which are covered, when receptive, with a copious liquid exudate is regarded as 'wet' (e.g. Orchidaceae, Fabaceaee, and Ericaceae), and if it has only an adhesive coating or pellicle it is classified as 'dry' (e.g. Poaceae, Brassicaceae, and Boraginaceae; refs 14 and 15); several tests to reveal the stigma receptivity (*Protocols 14–17* and *Table 4*) are based on this fact.

The exudate and secretions of the stigmas and style may fulfil several roles in the pollen–stigma interaction, such as:

- control of pollen adhesion, hydration, and germination (9)

61

Table 3. Main features of bi- and tricellular pollen grains

Feature	Bicellular pollen	Tricellular pollen
Viability retaining	Often tolerates prolonged storage	Loses viability shortly after maturity
Germination *in vitro*	Germinates readily in culture	Difficult to germinate in culture
Respiration rate	Low	High
Response to incompatible pollination	Arrested at the style	Arrested on the stigmatic surface
Self-incompatibility system	Gametophytic	Sporophytic
Growing on compatible style	Two phases of growth	One phase of growth

Sources: refs 1, 8–12.

Table 4. Methods for estimating stigma receptivity

Method	Advantages	Disadvantages
Benzidine + H_2O_2	Easy to carry out, also for dry stigmas	Not quantitative. Benzidine is highly carcinogenic
H_2O_2	Easy to carry out, quick results; suitable for field conditions	Not quantitative
Peroxidase test paper	Handy, easy to carry out, immediate results	Not suitable for dry stigmas. Not quantitative
Controlled pollination	The most accurate method, quantitative	Requires a substantial amount of fastidious work and entails a lengthy procedure
Alpha-naphthyl acetate test	Simple. Reliable results	Damaged or cut stigmas will give positive reaction although not receptive
I. Baker's test	Not complicated. Reliable results. Quantitative	Some pollen may germinate even if the stigma is not receptive

- protection from predators and microbial attack (16) and to prevent stigma dehydration (17)
- nutrition of the pollen grain during its growth (18)
- reward to pollinators (19)

2. Assessments of pollen viability and stainability

Pollen viability tests are used for:

- evaluation of the fertility of a given parent plant (particularly in breeding programmes)
- monitoring the pollen state as a function of storage conditions
- evaluation of the chance of pollen germination in studies of pollen–stigma interaction
- studies of incompatibility and fertility
- spotting individual hybrids in wild populations
- evaluation of the chances of pollen germination after exposure to various conditions, dispersal by agents, etc.

Protocol 1. Stainability of pollen grains

Materials
- vital dye: methylene blue (1%); neutral red (1%); aniline blue (1%)

Method
1. Put a sample of pollen grains in a droplet of the dye and cover with a slip.
2. Replace the dye solution with water or glycerol after 5 min.
3. Count the percentage of the dyed pollen grains.

Pollen stainability may depart considerably from the real value of pollen viability. Thus it should be used with a caution or only as a preliminary or complementary test (20).

The staining capacity depends not on the viability but on the content of pollen grains. Even herbarium specimens may show a positive reaction (21).

Protocol 2. Staining of aborted and non-aborted pollen—Alexander's procedure (modified from ref. 22)

Materials
- 10 ml ethanol 35%
- 10 ml malachite green (1 ml of 1% solution in 95% ethanol)
- 50 ml distilled water
- 25 ml glycerol
- 5 g phenol

Protocol 2. *Continued*

- 5 g chloral hydrate
- 50 ml fuchsin (5 ml of 1% solution in water)
- 5 ml orange G (0.5 ml of 1% solution in water
- 1–4 ml glacial acetic acid

Method

Mix all the ingredients in the order of the above list and store in a dark bottle at room temperature. Renew the mixture after a month. The amount of the glacial acetic acid depends on the thickness of the pollen grain wall, 1 ml for thin-walled pollen, and up to 4 ml for thick-walled pollen.

A. *Staining of thin-walled pollen*

1. Mount pollen samples directly into a drop of stain, cover with a coverslip, warm over a small flame, and examine through a microscope.
2. Keep the stained sample at 50 °C for 24 h if the differentiation between aborted and non-aborted pollen is not clearly satisfactory. Aborted pollen stains green and non-aborted stains red.
3. Seal the mount for further examination.

B. *Staining of thick-walled pollen*

(i) *Non-sticky pollen*

1. Add 3 ml glacial acetic acid to 100 ml of the stain.
2. Cover the sample of pollen with an excess of the stain mixture (e.g. in grooved glass slide or small tube).
3. Incubate at 50 °C for 24–48 h.
4. Examine microscopically; if the differential staining is not sufficient, add 1 drop of 45% acetic acid to the staining mixture and mix gently.
5. Prepare pollen mount and seal it.

(ii) *Sticky and oily pollen*

1. Add 3 ml acetic glacial acetic acid to 100 ml of the stain and mix.
2. Fix mature but non-dehisced anther for 24 h in Carnoy's fixative (3 ethanol:2 chloroform:1 glacial acetic acid) to remove the pollen-kit (it will improve the staining).
3. Transfer through an alcohol–water series to water (70, 50, 30% ethanol, 30 min in each concentration, and finally rinse in water).
4. Dry the anther with tissue paper.
5. Split the theca walls in a droplet of the stain, release the pollen, then remove the anther remains.

6. Cover the droplet with a coverslip and incubate in 50 °C for 24 h.

7. Add 1–2 drops of stain after 24 h under the coverslip to refill the evaported stain.

8. Examine the slide after 24 h. If the staining is too green, add 1–2 drops of glacial acetic acid and repeat steps 5 and 6 of this section.

The above test is too positive for assessing germinability of pollen grains since the demonstration of the presence of cytoplasm *per se* is not a guarantee that pollen grains are viable and able to germinate.

Protocol 3. The fluorochromatic reaction (FCR) test for pollen viability (ref. 12, p. 298; refs 20 and 23)

Materials

- 20 mg of fluorescein diacetate in 10 ml acetone
- 15% sucrose solution

Method

1. Preparation of the reagent:
   (a) Place 10 ml of the freshly made sucrose solution in a transparent vial.
   (b) Add the fluorescein diacetate in acetone drop by drop (about 1–3 drops) until it turns a light, milky, or greyish smoke colour.

2. Store dehydrated pollen grains for 10–30 min under high relative humidity before the test to enable membrane recovery and to prevent false-negative results.

3. Check in advance that there is no pollen bursting in this particular concentration of sucrose. In such a case use 20 to 30% sucrose.

4. Disperse the pollen sample in a drop of fluorescein diacetate. Put the slide in a Petri dish lined with wet filter paper for 10 min and cover the drop with a slip.

5. Examine the drop under a fluorescent microscope. Use a violet exciter filter which emits a beam of purple-blue light.

6. Score pollen grains with bright golden-yellow fluorescence as viable. Empty, undeveloped grains will not fluoresce at all. Grains losing their viability will have a dim 'ghostlike' fluorescence compared to the bright, fully viable grains.

7. To record the viability percentage, count the grains under white light first and then switch to the fluorescent light to see how many grains remain visible.

Protocol 3. *Continued*

8. If the exine itself shows bright fluorescence, compare the fluorescence of fresh pollen grains to killed grains (by exposure to *c*. 80 °C for 2–12 h).

Note

Remember, lipids are autofluorescent under the blue-purple light so even dead, empty grains will show some weak, yellowish fluorescence from the pollen-kit droplets. Observations and grain counts must be completed within 10 min after placing the coverslip over the specimen. Fluorescein diacetate in sucrose solution must be made up fresh each time.

This seems to be the most reliable method to evaluate pollen viability although it is a test for active esterases and integrity of the plasmalemma of the vegetative cell and *not* a test of viability. Pollen grains subjected to desiccation may give negative reactions but positive after recovery (20).

Protocol 4. Tetrazolium chloride test for pollen viability (20, 24, 25)

Materials

0.5%, 2,3, 5-triphenyl tetrazolium chloride (TTC) in the optimum sucrose solution for the pollen germination of the species (see *Protocol 7*). Keep the solution in a dark bottle.

Method

1. Put a sample of pollen in a drop of the medium and cover immediately to exclude oxygen, which can inhibit dye reduction.

2. Put the slide in a Petri dish lined with wet filter paper.

3. Keep the preparation at 50 °C in the dark for up to 2 h.

4. Score only the red-coloured pollen grains which are in the central area of the sample.

5. Treat the test as 'not accurate' if more than 20% of the pollen grain sample gives intermediate results (e.g. light to very light red or pink) and try direct pollen germination or the FCR test.

The intensity of the pollen staining decreases towards the margin of the cover-slip and in the region of air bubbles. To minimize sources of error, only pollen grains remote from the faded, high oxygen area should be counted (24). Desiccated pollen grains may give false-positive reactions (20).

Protocol 5. Enzymatic examination of pollen viability (Irene Baker pers. comm. and ref. 26)

Materials

Substrate for alcohol dehydrogenase:

- 0.1 M phosphate buffer; pH 7.3–7.5. Dilute 1 part buffer to 2 parts distilled water, 10 ml.
- nitroblue-tetrazolium (NBT)—just enough to give a slight yellow colour
- nicotinamide adenine dinucleotide (NAD) 6 mg
- ethanol 95% 500–1000 μl

Method

1. Prepare a substrate solution by mixing all the ingredients in advance and keep in a small vial.
2. Put a fresh pollen sample on a slide and add 20–30 μl of the substrate.
3. Mix the pollen sample carefully with the substrate. Keep the slide for 15–20 min in a closed Petri dish (with a wet filter paper at the bottom) to prevent drying and/or contamination.
4. Remove the cover, and allow the pollen and the substrate to dry under room conditions.
5. After 30–60 min, cover the dried spot with glycerol-jelly or a permanent mountant (e.g. polyvinyl lactophenol *Protocol 11*). Development of a dark-blue colour around the pollen grain indicates its viability.
6. Express the estimated viable pollen as the percentage of pollen around which a blue colour was developed.

Note

The same procedure can be applied to check stigma receptivity, after which the stigma can then be applied to the substrate instead of pollen. By changing the substrate, energy source, and dye, the activity of other enzymes, such as malate dehydrogenase, succinate dehydrogenase, α-glycerol-phosphate de-hydrogenase and non-specific esterase, can be tested. The tests are universally applicable, but the presence of oily pollen-kit needs a vigorous stirring of the pollen–substrate mixture (26).

Protocol 6. The peroxidase reaction as an indicator of pollen viability (27)

Materials

- 10 ml agar 2%
- 5 ml H_2O_2 3%

Protocol 6. *Continued*

- 1.8 ml benzidine 1% in ethanol 60%
- prepare the medium within 1 h of its use

Method

1. Mix the pollen with several droplets of the test medium. Watch under the dissecting microscope. Appearance of oxygen bubbles, accompanied by the oxidation of the benzidine, and colouring of the pollen grains in blue, indicate pollen viability.
2. Express viability estimation as the percentage of the deeply blue pollen grains in several samples each of 100 pollen grains.
3. Carry out the examination 10–30 min after the pollen is introduced to the test medium. In some species, the maximum reaction is well-achieved after 5 min, be aware of this and check at 5-min intervals.

Caution: Benzidine is highly carcinogenic and discarding this method is recommended, if possible. Another drawback is that there is a need to calibrate the results to pollen germination rates *in vitro*. Estimation of pollen viability based on benzidine tests may not be correlated to seed set (28).

3. Pollen germination

Many pollen grains can germinate in water or aqueous solutions of sucrose with no additives although the pollen of some species (such as tricellular pollen grains) needs a special substrate for germination. Germination of pollen grains in increased sucrose solutions may serve to evaluate the osmotic relations of the pollen grains, and to find the optimal concentration for germinability tests for each species (it ranges between 0 to 50% sucrose).

It should be noted that germination success in sucrose medium may depend on the humidity to which the pollen grains were exposed prior to the germination test, and the pollen age (3). To minimize these effects, and to standardize the experiments, pollen should be extracted directly from the anther to the experimental solution as soon as possible.

Protocol 7. Pollen germination rate in sucrose solutions

Materials

- 3 cm Petri dishes
- sucrose solutions (0, 5, 10, 20, 30, 40, 50, 60%) as percentage by weight (g sucrose/100 g solution)
- 2×10^{-3} M H_3BO_3

- 6×10^{-3} M $Ca(NO_3)_2$
- hand refractometer
- Vaseline
- methylene blue

Method

1. Prepare the suitable sucrose concentration stock in a mixture of 50% H_3BO_3 and 50% $Ca(NO_3)_2$ by volumes. Check the sucrose concentration in a refractometer prior to the experiment.

2. Mark 4 × 4 squares on the inner side of a 3-cm Petri-dish cover (number the grids prior to adding of the droplets).

3. Put 10 μl of the tested solution on to the centre of each square.

4. Add fresh pollen to each droplet (*note*: not too much—about 200 grains per droplet). Each row can be used for a different plant/population/ species.

5. In the bottom plate put 2 ml of solution of the same sucrose concentration as that which is on the inner side of the top plate.

6. Smear the edges of the bottom plate with a generous layer of Vaseline.

7. Turn the cover gently (the droplets will stay in their original locations).

8. Repeat the procedure for each sucrose concentration.

9. Leave the plates (at least six replicates for each treatment) under room conditions for 24 h.

10. At the end of the experiment, put all the dishes in the refrigerator (± 0 °C) until the examination (which is time-consuming!).

11. To check the germination percentage, open the dish cover and add a little droplet of methylene blue to each sample. Then transfer each sample to a slide, cover gently, and count the pollen grains for germination rates, or mount the samples temporarily with nail varnish to prevent evaporation (they will keep for several days under room conditions) until counting.

12. If the experiment examines the germination dynamic vs. time, leave the plates at room conditions. Check the percentage of pollen germination periodically under the microscope via the closed cover plates or freeze replicates at every given period of time and count later.

13. Express the results as a percentage of germination vs. concentration for each examined time-period.

14. In germination rate experiments, draw the germination course vs. time. Generally, an S-shaped curve is obtained. To compare treatments, species, etc., use the following parameters:

  M the final percentage of germination

Protocol 7. *Continued*

    S the start of the germination (of a given species in a given concentration); the time needed to reach a value of 1/6 of the final germination (M) in that particular concentration

    R rate of germination, 4/6 of the final germination (M) divided by the time needed for germination from 1/6 M to 5/6 M

The optimum sucrose solution for pollen germination can be used to evaluate the maximal pollen germination rate as an indicator of pollen viability. Heslop-Harrison *et al.* (20) mentioned that the rate of pollen germination *in vitro* depends largely on the experimenter's success in finding the optimal medium, and this has to be considered, while counting germination as a criterion for viability. Especially in tricellular pollen grains there are difficulties in getting high germination rates, and many different media were suggested for various species (see ref. 12 for review).

Freshly shed pollen of some species shows low germinability, the maximum being obtained only after it has been kept for a period under high humidity (e.g. 30 min at 95% RH (Relative Humidity). After pollen is 'conditioned', high germination is achieved (3, 29).

Pollen grains cultured in dense and large populations germinate better and form longer pollen tubes than pollen germinated in small and evenly distributed populations (30).

Pollen germination rate *in vitro* may be low, but can produce satisfactory fruit set *in vivo* and vice versa (21). Thus *in vitro* results of pollen germination do not necessarily count for pollen capacity to fertilize ovules or to produce seeds.

The same procedure can be used to check the pollen germination in salt solutions (e.g. of sea-shore plants) or any other liquid. For additional *in vitro* tests and specific requirements for pollen germination, see refs 12 and 25. This method (*Protocol 7*) enables the minimization of osmotic changes in the solution and the simultaneous testing of large samples. It also enables the examination of each sample gradually later on and the following of the germination rate (the samples can also be photographed periodically) without any intervention.

Any opening of the Petri dish for manipulation may cause a drastic change of the micro-environment around the droplets and thus changes in their concentration may occur. Differential germination rates in the central zone of the droplet, in comparison to the margin, are overcome by an even dispersion of the whole droplet at the final counting.

Sources for intraspecific variation in pollen germination and viability tests:

- pollen dehydration and rehydration states (3, 29)
- pollen age (31, 32)
- weather effect (29, 33)

- location of the flowers on the canopy (34)
- genetic variation between individuals (30)
- timing at the season (35)
- exposure to high temperature (36)
- hybrids and introgressive populations may have high rate of aborted pollen

4. Pollen histochemistry

Pollen histochemical analyses are carried out for the following reasons:

- possible relation between the pollen content and the mode of pollination
- study of pollinator foraging behaviour, nutritional demands, and pollination mode
- pollen content and composition in relation to phylogeny

4.1 Starch in pollen grains (37)

Method: Immerse the pollen sample into the IKI solution (see Appendix A2) and examine it under the microscope. A dark bluish-black colour indicates the presence of starch. New starch may appear as purple.

Generally, it is recommended to use only fresh and mature pollen grains for starch presence; normally starchless pollen may contain starch when premature (38). It is better to use already shed pollen to ensure maturity. At least in Araceae, dried herbarium specimens yielded reliable positive data for the starch presence survey (39).

Baker and Baker (40) have shown that starchless pollen grains (which have a lot of lipids) are typical of bee-pollination, particularly where no other reward is offered by the flower, and also of fly-pollination. Starchy grains (which have some lipids) are typical of species that are self-pollinated, wind-pollinated, pollinated by Lepidoptera or by birds.

4.2 Lipids in pollen grains (45)

Method: Immerse the pollen sample in a drop of Sudan IV (see Appendix A2). A red colour indicates the presence of lipids. Use only freshly made stocks of Sudan IV and examine the sample within 2–3 min after the dye application. Sometimes the red coloration disappears after 5–10 min.

Baker and Baker (40) have shown that starch-rich pollen is lipid-poor, and vice versa, so both tests for starch and for lipids are checked simultaneously.

5. Pollen fixation and preservation

Pollen is fixed and preserved for the following purposes:

- study of the pollen grain amount and diversity carried by a potential pollinator

- pollen reference collection
- presence of pollen grains on the stigma (e.g. conspecific vs. foreign pollen grains)
- preserved pollen is used when fresh pollen is not available, especially in crop breeding

5.1 Freezing pollen

Pollen, like the sperm cells of animals, can be frozen for use in experimental crosses, as flowers, isolated by different seasons, become unavailable. Put dehiscent anthers at the base of a small, transparent vial that has been laid on its side. Place one or two crystals of silica gel at the mouth of the vial but do not allow them to touch the anthers. Seal the vial and leave it for 6 h or overnight. Remove and discard the silica gel crystals. Seal the jar and tap it against a table to shake pollen grains from the anthers. Remove the anthers from the vial and place them in another vial and repeat the shaking and banging process until the interior sides of several vials have been dusted with pollen grains. Label the vials and store in a freezer at −4 °C for best results. Specimens may last as long as 6 months. Always test some of the grains in a freezer vial for viability before making a cross. Once the vial is opened it should not be resealed and returned to the freezer (P. Bernhardt, pers. comm.).

Protocol 8. Pollen staining and fixation with gelatine–fuchsin (42)

Materials

- 175 ml distilled water
- 150 ml glycerol
- 50 g gelatine
- 5 mg crystalline phenol
- basic fuchsin crystals

Method

1. Add the gelatine to the distilled water in a large beaker and heat until dissolved. Then add the glycerol and the phenol and mix gently.
2. Add some crystals of basic fuchsin and mix until the solution is pale pink to pink. Add the crystals gradually, and stir gently while the solution is still warm.
3. Filter the solution through glass wool and store it in clear uncontaminated containers.
4. When needed, take a small cube of the solid gelatine–fuchsin, melt it gently, add the examined pollen sample and cover.

5. When the preparation is ready it can be kept in a cool and dark place for 3–4 years.

Note

Do not overheat the melted gelatine. It could spoil the mountant. Take a small cube (2 × 2 × 2 mm) of the gelatine to prevent flow below the cover-slip. Under field conditions, an alcohol heater (or even a single match) is very useful and is fully satisfactory. This technique can be applied to a whole stigma (to evaluate the pollen load) or directly to an insect or other animal to identify and count its pollen load.

Protocol 9. Pollen staining and fixation; modified Calberla's solution (ref. 43, and P. Bernhardt, pers. comm.)

Materials

- 5 ml glycerol
- 10 ml 95% ethanol
- 15 ml distilled water
- crystals of aqueous basic fuchsin (mixed with enough water to make a thick slurry or supersaturated solution)
- melted glycerol-jelly (see Appendix A2) (optional)

A. *Preparation of the Calberla's fluid*

1. Mix the first three ingredients. Filter the supersaturated solution of aqueous basic fuchsin and reserve the dark purple liquid. Add the purple liquid drop by drop to the solution until it turns a pale, transparent pink. Do not allow the stain to become a dark ruby or claret red!

2. Add 2–3 drops of the melted glycerol-jelly. This may dilute the basic fuchsin to the point where more basic fuchsin must be added to return the pale pink tint. Store at room temperature.

B. *Use*

1. Place 2–3 drops of the Calberla's solution directly on the top of the pollen grains and wait 7–10 min *before* adding the cover-slip.

2. Most pollen can be identified after adding the cover-slip or it can wait until the following day.

3. Glycerol-jelly causes the solution to act as a semipermanent mount. You can seal the edges with nail varnish and store the slide in a cool dark place for 3–4 years.

Protocol 9. *Continued*

4. To remove pollen from a dead insect 'wash' it gently with 2–3 drops of absolute ethanol, remove the insect, let the ethanol evaporate, and place the Calberla's solution as in step 1.

5. Viewing of the slide is best 25 h later when the colouring is deep enough to recognize details. The wall of the pollen is stained in pink-red, illuminating the pores, colpi, and exine sculpture. The pollen-kit will not be stained and its contrasting pigmentation will be seen on the exine surface as a series of oil droplets.

If the solution is too red it will stain both the pollen and the cytoplasm. That will obscure pores, colpi, and opercula which are needed for final identification. A light pink solution stains *only* layers of the exine and tectate sculptures. Because Calberla's fluid contains water, the grains will hydrate and pollen may appear distorted when compared with the way the grain looks under the SEM.

This method is especially convenient for quick fixation in the field for reference collection, pollen brushed off from foragers, or squash of stigmas.

Protocol 10. Pollen fixation in polyvinyl–alcohol (44)

Materials

- glass rod
- Petri dishes (3 or 5 cm)
- 12.5 g polyvinyl–alcohol (e.g. Airvol 240, sold as a powder)
- 100 ml distilled water
- water bath
- oven

Preparation

Dissolve the polyvinyl–alcohol, with constant stirring, in 100 ml distilled water in a beaker in an 80 °C water bath for 30 min.

Method

1. Prepare the pollen sample (fresh or after acetolysis, *Protocol 12*) in 5 ml distilled water (or 5% sucrose if pollen burst occurs in the distilled water).

2. Pour the pollen sample into a Petri dish and add 1.5 ml of the dissolved polyvinyl–alcohol solution and mix well with a glass rod to spread it evenly all over the dish surface.

3. Be careful that the suspension does not creep up the wall of the Petri dish.

4. Dry the plate at 60 °C for 1.5 h.

5. Pull out the film gently with large forceps.

6. With a cork hose, punch several discs of known area from various parts of the film and mount them on a glass slide under a coverslip and count the pollen grains in each sample.

7. Calculate the average number of pollen grains per square centimetre and multiply with the film average area.

The whole film can be kept as one sample and examined under the microscope, especially when pollen density is low. Total pollen number per cubic centimetre is

$$\frac{\text{surface of the film (cm}^2)}{\text{area of counted part (cm}^2)} \times \frac{\text{pollen counted}}{\text{volume of the sample (cm}^3)}$$

The films are easy to store in envelopes, for any further examination or identification.

Irene Baker (pers. comm.) developed a polyvinyl–alcohol mountant (Appendix 2). This clear mountant is kept in a dark bottle. It can be used with dyes which are stable in the acid medium such as iodine crystals and cotton blue. This allows for staining and making slides permanent at the same time. If the mountant shrinks on drying, more of the medium may need to be added as it dries.

5.2 Estimation of pollen grain number per flower

Pollen grains are counted for the following purposes:

- pollen:ovule (P:O) ratio determination (see Chapter 2, Section 4)
- in gene flow studies
- in comparison of wind vs. animal pollination syndromes
- in studies of pollination efficiency and dynamics of pollen dispersal vs. time or pollinator activity

---

Protocol 11. Ultrasonic pollen removal (ref. 45, pp. 34–7)

Materials

- acetone
- sonicator (e.g. W-220G, Heat System–Ultrasonics, Farmingdale) with 13 mm diam. horn

Protocol 11. *Continued*

- 15 ml vial (borosilicate glass shell vial with polyethylene snap cup, model 60965. Kimble, Vinhead, NY)

Method

1. Put the specimen (anther, stigma, or bee) from which the pollen should be removed, in a vial with 8 ml acetone for 30 min.
2. Insert the horn (see *Figure 1*) up to a height of 8 mm above the bottom of the vial. Operate the sonicator for 30 sec at an amplitude of 60 μm.
3. Rinse the pollen grains clinging inside by a small jet of clean acetone from a squeeze-bottle when pouring the suspension for counting.
4. Stain (see *Protocol 9*), mount (see *Protocol 10*), or count (see Chapter 2, *Protocol 3*) the pollen grains.

Vaissiére (45) developed the procedure for cotton stigmas; the exact sonicating duration (20–100 sec), amplitude (36–84 μm), and distance between the tip of the horn and the bottom of the vial (8–29 mm) have to be worked out for each plant (stigma or anther) or insect.

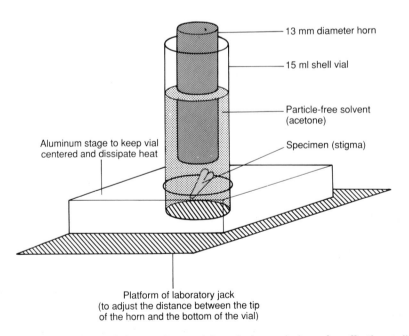

Figure 1. Arrangement of the specimen, vial, and ultrasonic horn for effective pollen removal from stigmas, bees, or anthers. (From ref. 48, used with the author's permission.)

Protocol 12. Acetolysis of pollen grains (ref. 46, pp. 79–80; ref. 47, p. 44)

Materials

- glacial acetic acid
- anhydrous acetic acid
- 95% sulfuric acid
- fume cupboard
- centrifuge
- water bath

Method

1. Put the pollen grain sample in glacial acetic acid for 10 min. Centrifuge and decant; discard the supernatant.

2. Remove the *deposit* into 6 ml of freshly made mixture of anhydride acetic acid and concentrated sulfuric acid (9:1 v/v). The mixture is highly corrosive and reacts vigorously with water. Care should be taken in its handling, especially in ensuring that materials with which they come into contact are dry. Heat gently to the boiling point in a water bath for 3 min, stirring continuously with a glass rod. Centrifuge as above, and decant the supernatant carefully into running water.

3. Resuspend in glacial acetic, centrifuge, and decant.

4. Resuspend in distilled water, centrifuge, and decant the supernatant. Repeat.

5. Embed the sample on a microscopic slide with polyvinyl lactophenol (*Protocol 10*) or gelatine-fuchsin (*Protocol 8*).

Caution

- The sulfuric acid should be added drop by drop to the anhydrous acetic acid and stirred.
- The whole procedure should be performed in a fume cupboard with gloves and eye protection.
- Dispose of surplus acetolysis mixture by pouring it very carefully into cold water.

Acetolysis certainly has considerable influence on the final form of the colpus and porus, depending upon the time for which the mixture was boiled and the precise composition of the mixture. Even slight departure from the

9:1 ratio of anhydrous acetic acid to sulfuric acid can result in structural changes in the pollen grain (ref. 47, p. 47).

Acetolysis destroys all the pollen material with the exception of sporopollenin that forms the outer pollen wall, the exine. The removal of the pollen-kit of non-sporopollenin pollen-connecting threads, by acetolysis, may distort the functional interpretation of such pollen types (48).

Wind-pollinated trees and grass pollen are easily over-acetolysed. Over-acetolysed grains are often crumpled and distorted and structural information may be lost (ref. 47, p. 57).

6. Stigma receptivity

Stigma receptivity is a crucial stage in the maturation of the flower which may greatly influence the rate of self-pollination, pollination success at different stages in the flower life-cycle, the relative importance of various pollinators, the interference between male and female functions, the rate of competition via improper pollen transfer, and the chances of gametophytic selection (49).

Any success in breeding experiments or artificial pollination procedures should be accompanied by tests on the timing and duration of the stigma's receptivity (*Table 4*). The duration of stigma's receptivity varies from a few hours up to 10 days. The age of the flower, the time in the day, and the presence or absence of stigmatic exudate, may all influence receptivity (4).

Stigma's receptivity can be determined experimentally by:

- the morphological changes, especially the presence of exudate (in wet stigmas) and colour changes (if there are any)
- determination of pollen germination and/or tube growth and/or seed set after pollination at different times relative to the flower opening
- staining or testing for esterase presence (or other enzymatic activity)

6.1 Applications of stigma receptivity tests

Stigma receptivity is studied for the following purposes:

- to identify the optimum flower age for artificial pollination procedures
- in studies of pollination efficiency, since transfer of pollen to pre- or post-receptive stigmas is a waste of pollen
- for studies of possible interference between the male and female functions
- for studies of breeding systems
- to determine relative effectiveness of pollinator visits made to flowers during different stages of anthesis

Protocol 13. Estimation of the stigma's receptivity with benzidine and H_2O_2 (refs 51 and 51 for H_2O_2 only)

Caution: Benzidine is highly carcinogenic and should be used with utmost care.

Materials

- 1% benzidine in 60% ethanol, hydrogen peroxide (3%), and water, 4:11:22 by volume.

Method

1. Excise the pistil and immerse it in a depression of a slide with the reaction solution.
2. If the stigma is receptive (indicated by the presence of peroxidase), it breaks the hydrogen peroxide and the oxidation of the benzidine gives a blue colour.
3. In a simpler version, it is enough to immerse the stigma into hydrogen peroxide (3%) and to observe the bubbles. This method is less accurate but is very useful especially under field conditions.

Protocol 14. Determination of stigma receptivity *in vivo* by artificial cross-pollination

Materials

- fine mesh nets
- coloured or numbered tags

Method

1. To ensure stigmas are virgin, cover flowers with fine mesh nets before the experiment. If there is any possibility of self-pollination, the flower must be emasculated in advance.
2. Deposit conspecific pollen from another plant (ramets, clone, etc.) on a sample of virgin stigmas at various times during the experimentation; a different sample each time. Mark each flower in each treatment clearly with coloured or numbered tags.
3. During all of the procedures, keep the flowers under the nets to prevent any other pollen transfer.
4. After a certain period (2–4 h, depending on the species and the experiment schedule), check the pollinated stigmas for germination of pollen grains (*Protocol 19*).

Protocol 14. *Continued*

5. If the stigmas remain fresh for several days, repeat the experiment 2–3 times a day. In any case, repeat the procedure throughout the stigmatic life-cycle, from buds to withering. The percentage of the germinating pollen grains is used to express the state of the stigma receptivity.

Note

Use only fresh compatible pollen from newly exposed anthers, preferably full anthers that were kept intact under nets. Apply the pollen with the same fine brush to ensure the same quantity, more or less, of pollen in each repetition.

Protocol 15. Evaluating stigma receptivity with peroxidase test paper (52, 53)

Materials

● peroxidase test paper[a] PEROXTESMO KO (Macherey-Nagel, Cat. No. 90606).

Method

1. Gently press the stigmatic surface on a piece of the test paper: be careful not to squash or hurt it. If the stigma is too small carry out the test under a magnifier or dissecting microscope. A dark-blue colour shows the presence of peroxidase as an indicator of the stigma receptivity. Handle the paper with forceps, not with your fingers.

2. Repeat the examination at different stages of the flower development.

3. In protandrous species emasculate the flowers before anthesis to prevent contamination of the stigma with pollen.

4. Before every test, check for presence of pollen on the stigma.

[a] Use only fresh test paper, in 6 months it develops a brownish coloration which distorts the read-off of the blue colour.

This test is appropriate only for 'wet' stigmas which are characterized by making a detectable secretion on dry paper while they are receptive (for the taxonomical distribution of 'wet' and 'dry' stigmas, see ref. 54. This test is useful mainly in Orchidaceae, Fabaceae, and Ericales.

Protocol 16. Alpha-naphthyl acetate test for the presence of esterase activity (ref. 55, and P. Bernhardt, pers. comm.)

Materials

● 10 mg of α-naphthyl acetate powder (store in refrigerator after opening bottle)

- 0.25 ml acetone
- 20 ml of 0.1 M phosphate buffer (pH 7.4)
- 50–100 mg of fast blue B salt

Preparation of the reagent

1. Dissolve the α-naphthyl acetate powder in the acetone in a vial that will hold more than 20 ml of fluid. Add the phosphate buffer.
2. Shake the tightly stoppered vial for about 10 min or until the initial 'milky' colour of the fluid begins to break up.
3. Add the fast blue B salt and shake so that everything is well mixed.
4. To make a large amount of the phosphate buffer for use over a period of weeks or months it is suggested that you dissolve 3.12 g of $NaH_2PO_4 \cdot 4$ H_2O in 19 ml of distilled water and then dissolve 3.58 g of $Na_2HPO_4 \cdot 7$ H_2O in 81 ml of water. Combine both solutions. Bottle and store on the lowest shelf on the refrigerator. (Buffer may crystallize under low temperatures and crystals must be dissolved before attempting the esterase test by immersing buffer bottle in a warm water bath.)

Method

1. Filter the stain and apply it directly on a selection of stigmas, taken at different stages of the flower life-span, until they are completely immersed, for 2–5 min.
2. Wash the stigmas in distilled water before observing under a dissecting microscope. If the stigmatic surface has produced esterase (a sign that the stigma is now receptive to pollen) the stigmatic secretions and/or the stigmatic papillae will turn a deep blue-black.
3. To determine the depth of the esterase production (indicated by the production of the blue-black colour) retain the stigmas for sectioning, using a freezing microtome.

The α-naphthyl acetate powder and acetone solution may be made in large quantities and drawn out with a micropipette as needed, but the solution and powder must always be stored in the refrigerator. If the solution is taken into the field it should be in some sort of container with ice-packs or dry ice. The solution mixed with fast blue B salt must be used immediately and cannot be saved. The stain reacts positively with virtually any enzymatic material so it will also stain the cut portion of the style, and it will always stain the stigma if the surface has been bruised or punctured before it has become receptive to pollen. Make sure the stigmas you test are undamaged by insects or the collection process. This is a water soluble stain for enzymes so it will usually fade away if you try to preserve the stigmas in any dehydrating fixative for any microtome work other than freezing techniques.

Protocol 17. Test for stigma receptivity (Irene Baker, pers. comm.)

Materials

- 300 mg agar
- 100 ml distilled water
- lacmoid solution, 0.1 N NH_4OH, 10 ml

Method

1. Dissolve, by boiling, the agar in the distilled water. Allow to cool to *c*. 45 °C.
2. Squash the stigma on a slide with a cutting needle, add some of the cooled agar to the maceration and mix well.
3. Wait 10–30 min for the compounds in the stigma to diffuse into the agar.
4. Add the tested pollen grains and mix them well in the agar.
5. Let the pollen germinate for several hours or overnight at room temperature. Put the slide in a covered Petri dish to prevent drying.
6. Add lacmoid solution to cover the slide for 30–60 min.
7. Wash off the stain with the diluted NH_4OH.
8. The callose of the germination tubes is stained blue.
9. Express the stigma state of receptivity as the percentage of the germinating pollen grains.

Check the stigmas at different stages of their development to determine the duration of receptivity and the optimum timing for maximal germination of pollen.

---

7. Stigma–pollen interaction

The presence of pollen grains on the stigma reflects only one facet of the reproductive cycle—the transfer of the male microgametophytes. Quantification of the pollen germination on the stigma will often supply the necessary evidence for the compatibility system (although later breakdown of SI can occur, ref. 7). The relation between the pollen load and number of germinating pollen grains and the number of these reaching ovules may indicate possible competition among the pollen grains (8). Pollen germination rate on the stigma vs. number of ovules can also evaluate the amount of pollen dose required to ensure full fertilization. This aspect is important, especially for determining pollination limits, seed production, and fecundity.

The identification of pollen grain presence on the stigma may help in studies of competition and clogging of the stigmas with an inappropriate pollen load. Pollen dose on the stigma is also used in studies of the pollination efficiency of various pollinators in a given period of stigma exposure.

7.1 Pollen load on the stigma

Estimation of the pollen load on the stigma is used as a criterion for 'pollinator intensity' (see *Protocols 8, 9,* and *10* for pollen staining and fixation). Germinating pollen grains on the stigma (*Protocols 18* and *19*) are used to evaluate pollen viability as well as the stigma's receptivity. The percentage of pollen germination on the stigma and the numbers of pollen tubes in the style may or may not account for fertilization rate (25).

Protocol 18. Detection of pollen tubes in the style (56–58)

Materials

- FPA solution (formalin 40%, concentrated propionic acid, 50% ethanol, 5:5:90 by volume)
- 8 N sodium hydroxide
- 0.1 N potassium acetate
- aniline blue
- fluorescence microscope with UV filter

Method

1. Fix the excised stigma and style in FPA for 24 h and then store in ethanol 70%. Wash in running tap-water before the next stage.
2. Soften the style for 5 h (depending on the species: the correct amount of time needed for softening can vary between 1 to 12 h) in sodium hydroxide. Rinse in tap-water for 1–3 h to remove the sodium hydroxide.
3. Stain with 0.1% aniline blue in potassium acetate for 4 h.
4. Squash the stained style under a cover-slip and observe under a fluorescence microscope equipped with a filter set (of maximum transmission 365 nm). Both pollen tube walls and the callose plugs should show a distinct bright yellow to yellow-green fluorescence.

Williams and Rouse (58) used a variation of this procedure which includes fixation in 1:3 acetic acid:ethanol (for 2 h); softening in an autoclave in 10% (w/v) of anhydrous sodium sulfite for 30–40 min at 104 kPa (15 p.s.i.); staining overnight in the dark in decolorized aniline blue stain 0.1% in 0.1 M K_3PO_4; each pistil or part of it being laid out straight on a microscopic slide and squashed gently in the stain. This procedure is applied especially to robust large stigmas.

Protocol 19. Staining germinating pollen and pollen tubes in the style (60)

Materials

- glacial acetic acid 45% and ethanol 70%, 3:1 by volume
- dye solution of 150 mg of safranin O and 20 mg aniline blue in 25 ml hot (60 °C) glacial acetic acid; filter before use

The mixture may be kept at room temperature for several weeks.

Method

1. Remove the style at the desired intervals after the pollination, and fix in ethanol–aectic acid for 1 h and then transfer to ethanol 70%.
2. Hydrolyse the style in 45% acetic acid at 60 °C for 10 to 60 min (until it becomes soft enough to be squashed, depending on the species).
3. Split the stigma longitudinally (under the microscope, best with micro-manipulator) with its strands or the lobes of the stigmatic tissues.
4. Stain the stigma or the strands for 5 to 15 min in the hot (60 °C) dye solution.
5. Place the stained tissue on a microscopic slide, add a drop of the dye solution, and gently squash the tissue under a coverslip. The slide is ready for examination.
6. Pollen grains are stained in blue, but the end of the pollen tubes are stained in red.

The staining solution may be kept at room temperature for several weeks. Excised styles can be kept for several days in acetic alcohol in the refrigerator. This method stains ungerminated pollen grains as well as pollen tubes in their early stages of germination. Staining of tubes in their later stages of growth is limited to the portion of the growing end. Unstained parts of the tubes can easily be discerned through their length, especially when oblique lighting is used. This procedure does not require a fluorescence microscope but is less sensitive than the former (*Protocol 18*) which stains the whole pollen tube walls. Sometimes the contrast between the pollen tube tips and the background is not clear enough and depends on the hydrolysis conditions.

8. Pollination quality

Pollen grains loaded on the pollinator have to pass several obstacles which reduce their chances of mating (*Table 5*). Some of the limitations are caused by pollinator dimensions and foraging behaviour in relation to stigma morphology and spatial positioning. The components involved in the fate of the pollen grain, from reaching the receptive target stigma to fertilization, have been studied mainly in relation to male fitness and sexual selection (see Chapter 2, *Figures 1* and *2*). But there should also be studies in relation to female function because the frequency of compatible pollen grains arriving on stigmas may be varied.

References

1. Knox, R. B. (1984). In *Encyclopedia of Plant Physiology*, Vol. 17, (ed. H. F. Linskens and J. Heslop-Harrison), p. 508. Springer-Verlag, Berlin.
2. Heslop-Harrison, J., Knox, B. B., Heslop-Harrison, Y., and Mattsson, O. (1975). In *Biology of the Male Gamete* (ed. J. C. Duckett and P. A. Racey), p. 189. Academic Press, London.
3. Heslop-Harrison, J. (1987). *Int. Rev. Cytol.*, **107**, 1.
4. Dumas, C., Knox, B. B., and Gaude, T. (1984). *Int. Rev. Cytol.*, **90**, 239.
5. Mulcahy, G. B. and Mulcahy, D. L. (1983). In *Pollen Biology and Implications for Plant Breeding* (ed. D. L. Mulcahy and E. Ottaviano), p. 29. Elsevier Bio-Medical, New York.
6. Gaude, T. and Dumas, C. (1987). *Int. Rev. Cytol.*, **107**, 333.
7. Seavey, S. R. and Bawa, S. K. (1986). *Bot. Rev.*, **52**, 195.
8. Mulcahy, D. L. and Mulcahy, G. B. (1987). *Am. Nat.* **75**, 44.
9. Hoekestra, F. A. and Bruinsma, J. (1975). *Physiol. Plant.*, **34**, 321.
10. Bar-Shalom, D. and Mattsson, O. (1977). *Bot. Tidsskr.*, **71**, 254.
11. Cereau-Larrival, M.-T. and Challe, J. (1986). In *Pollen and Spores: Form and Function* (ed. S. Blackmore and S. K. Ferguson), p. 151. Linn. Soc. Symp. Ser. 12. Academic Press, London.
12. Shivanna, K. R. and Johri, B. M. (1985). *The Angiosperm Pollen—Structure and Function*. Wiley Eastern, New Delhi.
13. Dumas, C. and Gaude, T. (1983). *Phytomorphology*, **3**, 191.
14. Heslop-Harrison, Y. and Shivanna, K. R. (1977). *Ann. Bot.*, **41**, 1233.
15. Mattson, O., Knox, R. B., Heslop-Harrison, J., and Heslop-Harrison, Y. (1974). *Nature (Lond.)*, **247**, 298.
17. Konar, R. N. and Linskens, H. F. (1966). *Planta*, **71**, 372.

Table 5. Components of the 'Pollination quality'

Pollination stage	Measurable parameter	Remarks
Pollen-loaded pollinator $\rightarrow$	No. of pollen grains attached to the pollinator	Heavily loaded pollinators are not necessarily the most efficient ones in loading the stigma with the proper pollen and vice versa. Some pollen may be out of reach stigmas, or else non-viable.
Flower visit $\rightarrow$	Amount/proportion of landing/entrance/touch of the flowers	Not all visitors are necessarily pollinators. There is a critical need to differentiate between visitors and pollinators. The most frequent pollinators may not be the most efficient in terms of pollen loading on the stigma.
Stigmatic contact $\rightarrow$	Amount/proportion of contacts between the pollinator's body and the stigmas	Contact with stigma does not necessarily means a pollination event even if the visitor is loaded with pollen. The Total Pollen Load (TPL) on the visitor's body has to be differentiated from the pollen on the contact area with the stigma, the Functional Pollen Load (FPL).
Pollen delivery to a receptive stigma $\rightarrow$	Amount/proportion of the stigmatic contacts resulting in pollen delivery, during the stigma's receptivity period	The pollen load on the pollinator is effective only in relation to chances of meeting the stigmatic surface. The number of pollen grains and their exact location on the pollinator's body are essential factors in relation to pollinator movement and spatial position of the stigma. Analysis of the number of pollen grains on the pollinators alone does not reveal anything about their chances of reaching the stigma.

Stage	Measurement	Notes
Number of pollen grains loaded on the stigma →	Number of pollen grains deposited on receptive stigma during one visit	In general, one visit of a pollinator may supply the needed amount of pollen. Foreign pollen may clog the stigma when 'interference competition' occurs.
	Number/proportion of conspecific compatible pollen grains of the whole pollen load during receptivity	May be dependent on the pollinator's flight distance, self-incompatibility, and the population structure system.
Pollen germination →	Number of germinating pollen grains	Pollen allelopathy or population effects may reduce the germination on the stigma, or 'mentor effect' may increase it.
Pollination tube growth →	The proportion of germinating pollen tubes	Competition between the germinating pollen grains may determine their success in reaching ovules.
Fertilization →	Percentage of fertilized ovules	Competition between fertilized ovules may result in abortion which may regulate final seed production.
Seed production	Proportion of seeds produced per flower and/or ovule	The product of the male reproductive success. Seed size and viability have also to be considered.

18. Loewus, F. and Labarca, C. (1973). In *Biogenesis of Plant Cell Wall Polysaccharides* (ed. F. Loewus), p. 175. Academic Press, New York.
19. Baker, H. G., Baker, I., and Opler, P. (1974). In *Pollination and Dispersal* (ed. N. B. M. Brantjes), p. 44. University of Nijmegen, The Netherlands.
16. Martin, F. W. and Brewbaker, J. L. (1971). In *Pollen, Development and Physiology* (ed. J. Heslop-Harrison), p. 262. Butterworths, London.
20. Heslop-Harrison, J., Heslop-Harrison, Y., and Shivanna, K. R. (1984). *Theor. Appl. Gen.*, **67**, 367.
21. Johri, B. M. and Vasil, I. K. (1961). *Bot. Rev.*, **27**, 325.
22. Alexander, M. P. (1969). *Stain Technol.*, **44**, 117.
23. Heslop-Harrison, J. and Heslop-Harrison, Y. (1970). *Stain Technol.*, **45**, 115.
24. Cook, A. and Stanley, R. G. (1961). *Silva. Genet.*, **9**, 134.
25. Stanley, R. G. and Linskens, H. F. (1974). *Pollen.* Springer-Verlag, Berlin.
26. Juncosa, A. M. and Webster, B. D. (1989). *Am. J. Bot.*, **76**, 59.
27. King, J. R. (1960). *Stain Technol.*, **35**, 225.
28. Janssen, A. W. and Hersmen, I. G. T. (1980). *Euphytica*, **27**, 577.
29. Corbet, S. A. (1990). *Isr. J. Bot.*, **37**, 13.
30. Vasil, I. K. (1987). *Int. Rev. Cyt.*, **107**, 127.
31. Palmer, M., Travis, J., and Antonovics, J. (1989). *Oecologia (Berl.)*, **78**, 321.
32. Dowding, P. (1987). *Int. Rev. Cytol.*, **107**, 421.
33. Smith-Huerta, N.L. and Vassek, F. C. (1984). *Am. J. Bot.*, **71**, 1183.
34. Oni, O. (1990), *For. Ecol. Manage.*, **37**, 259.
35. Eisikowitch, D., Kevan, P. G., Fowler, S., and Thomas, K. (1990). *Pollen Spores*, **29**, 121.
36. Struik, P. C., Doongist, M., and Boonman, I. G. (1986). *Neth. J. Agric. Sci.*, **34**, 469.
37. Jensen, W. A. (1962). *Botanical Histochemistry*, p. 201. Freeman, San Francisco.
38. Baker, H. G. and Baker, I. (1979). *Am. J. Bot.*, **66**, 591.
39. Grayum, M. H. (1985). *Am. J. Bot.* **72**, 1565.
40. Baker, H. G. and Baker, I. (1983). In *Pollen Biology and Implications for Plant Breeding* (ed. D. L. Mulcahy and E. Ottaviano), p. 43. Elsevier, Amsterdam.
41. Conn, H. J. (1961) *Biological Stains*, p. 76. Williams & Wilkins, Baltimore, Maryland.
42. Beattie, A. J. (1971). *Pan Pac. Entomol.*, **47**, 82.
43. Ogden, E. C., Raynor, G. S., Hayes, J. V., Lewis, D. M., and Haynes, J. H. (1974). *Manual for Sampling Airborne Pollen.* Hafner Press, New York.
44. Middeldorp, A. A. and Mijzen, P. (1986). *Pollen Spores*, **28**, 435.
45. Vaissiére, B. E. (1991). Honey bees, *Apis mellifera* L. (Hymenoptera: Apidae), as pollinators of upland cotton, *Gossypium hirsutum* L. (Malvaceae), for hybrid seed production. Ph.D. thesis, Texas A. and M. University, College Station.
46. Faegri, K. and Iversen, J. (1989). *Textbook of Pollen Analysis* (4th edn) (ed. K. Faegri, P. E. Kaland, and K. Krzyrinski). John Wiley, London.
47. Moore, P. O., Webb, J. A. and Collinson, M. E. (1991). *Pollen Analysis* (2nd edn). Blackwell Scientific Publications, Oxford.
48. Hesse, H. and Waha, M. (1989). *Pl. Syst. Evol.*, **163**, 147.
49. Galen, C., Zimmer, K. A., and Newport, M. E. (1987). *Evolution*, **41**, 593.
50. Galen, C. and Plowright, R. C. (1987). *Can J. Bot.*, **65**, 107.
51. Galen, C. and Kevan, P. G. (1980). *Am. Midl. Nat.*, **104**, 281.

52. Sullivan, J. R. (1984). *Am. J. Bot.* **71**, 815.
53. Koptur, S. (1988). *Syst. Bot.*, **8**, 354.
54. Heslop-Harrison, Y. (1981). *Nord. J. Bot.*, **2**, 401.
55. Pearse, A. G. E. (1972). *Histochemistry, Theoretical and Applied*, (2nd edn), p. 1303. Churchill Livingstone, Edinburgh.
56. Martin, F. W. (1959). *Stain Technol.*, **34**, 125.
57. Bhaduri, P. N. and Bhanja, P. K. (1962). *Stain Technol.*, **37**, 351.
58. Gurr, E. (1965). *Of Dyes in Biology*, p. 268–9. Leonard Hill, London.
59. Williams, E. G. and Rouse, J. L. (1990). *Sex. Plant Reprod.*, **3**, 7.
60. Dionne, L. A. and Spicer, P. B. (1958). *Stain Technol.*, **33**, 15.

4

Advertisement in flowers

1. Introduction

Floral advertisement includes several simultaneous stimuli: visual, olfactory, and short-range tactile cues. The magnitude of the advertisement depends on the flower colour, size, and shape, as well as the strength of the volatile emissions and their perception by the pollinators. Some pollinators may change their initial response to colours in the presence of scents (e.g. honeybees; ref. 1) and this fact should be kept in mind. Thus, a separate treatment of visual and olfactory cues is not realistic and the interrelationship between stimuli must also be considered or controlled in any study and experiment involving artificial or live flowers (see, for example, ref. 2).

For practical reasons, flower colour determination is considered separately from the measurement of flower size and shape. The chemical nature of the floral colours (3, 4) and colour vision of the potential pollinators (5, 6, 7) are beyond the scope of the present manual.

Under the general concept of 'advertisement', one may include all the stimuli which may attract visitors to the flowers. The various cues (visual, olfactory, and tactile) may act separately, but the summation of all of them, based on the background of the visitor's sensory perception, learning ability, and experience, determines the visitors' attraction and behaviour (*Figure 1*).

Although for the sake of discussion, there is a tendency to separate 'advertisement' and 'reward', the reward itself, such as yellow pollen (8) or UV-fluorescent nectar (9), (but see ref. 10) may often have a role in advertisement. If the advertisement is not matched with a reward, then it is considered a 'deception' (11). Advertisement is studied mainly in an attempt to identify 'pollination syndromes' that are typical of a specific group of pollinators (12) and to uncover their relation to pollinator behaviour (13) and contribution to the plant's reproductive success (14).

Several aspects may be defined concerning the analysis and understanding of the role of flower advertisement in pollination:

- analysis and measurement of stimuli in relation to floral morphology and field observations of flower visitors

- experiments with flower models in relation to pollinator behaviour in the field or laboratory (frequently with honey-bees and bumble-bees)
- manipulation of flower size, form, colour, density, and presence of other co-flowering species
- the influence of floral changes, variations, or polymorphism in advertisement on pollinators, and the resulting seed production, sexual selection, and pollen flow

2. Flower colours as advertisement

The visual stimulus is a combined effect of the target colour (and combination of colours and pattern), shape, and size (which determine the area of the reflecting surface), and the flicker stimulus which is the result of motion and the structure of the pollinator eye. The range of the visible spectrum of a given pollinator is the primary factor which enables the choice of flower colour to which it may be attracted (15).

When measuring and evaluating the possible ecological implications of floral colour, several issues should be borne in mind (16). Ultraviolet light is only one of the wavebands visible to insects, but its presence has little meaning without reference to the insect's visual spectrum and insect colour vision. The characterization of the flower colours should also include the colour contrast between the object and its background, as well as the spectral reflectance. Although it is debatable if there is an innate colour preference (13), several groups of pollinators show an innate responsive behaviour to specific colours (see refs 6 and 7).

Flower colour could be a result of absorption and/or reflection of several ranges of the spectrum including the UV wavelengths (15, 18). Visual advertisement in flowers is only one means by which flowers communicate unidirectionally with their pollinators. Flowers emit visual cues which supply the information on the specific identity of the flower's receptive stage and which release pollinator's behaviour involved in feeding and pollen transfer (19).

The timing of the advertisement usually coincides with pollen and/or reward presentation. Visits before or after the fertile period may damage the flower, and if rewards are not available, they may cause negative associative learning between the flower and its visitor. This, in turn, reduces their efficiency when acting as pollinators to other flowers of the same species. Colour change after pollination may reduce this effect (20).

Flower colour *per se*, is only one of several advertisement stimuli and should be interpreted in relation to other factors such as:

- the background colour and its reflectance
- colour contrast

- other simultaneous cues (odour, size, and shape)
- the presence and the exposure of reward
- other co-flowering species

Floral colours are analysed in various studies such as:

- the flower colour spectrum of a flora, geographic region, or ecosystem
- the seasonal shift of flower colours in a given locality and/or ecosystem
- flower colour in relation to a specific pollinator and its behaviour
- the intraspecific flower colour variability and its implications to pollinators, seed production, and gene flow
- colour change during the flower's life-cycle in relation to pollination and reproductive success

2.1 Measurement of flower colour

Flower colour is the result of the presence of chemical compounds as well as of the physical microstructure of the flower surface (21). Colour determination may be achieved by the following methods:

- human visual evaluation
- photography using different filters and a comparative grey-scale
- colorimetric measurements of the reflectance with a spectrophotometer
- direct visualization using a video camera

For the advantages and disadvantages of the various methods see *Table 1*.

Table 1. Methods of colour determination and measurement

Method	Advantages	Disadvantages
Human visual evaluation	A rapid method for large samples, data can be assembled from existing flora and records	Exposed to equivocal interpretation and definition, various authors use different criteria for colour naming
Colorimetric measurement	Gives accurate measurement of the reflectance over the whole spectrum	Limited to the laboratory. Requires expensive equipment
Photography with different filters	Gives accurate data on the reflectance, can be used in the field. Relatively inexpensive	Requires care (focus calibration, comparative grey-scale)
Video-viewing	Handy to use in the field. If used with a grey-scale and various filters, reflectance data can be acquired	Needs to separate UV emissions from AVS emissions

2.2 The determination of the flower colour by human vision

The human visual spectrum (400–700 nm) is different from that of animals (300–650 nm) especially in the UV range; thus, any colour determination should be accompanied with a parallel UV assessment. To prevent misunderstandings concerning colour names and ambiguous descriptions of hues, intensity, and transitional colours, an official colour chart (see, for example, ref. 22) should be used for the standardization of colour description in the human visual spectrum. Data are sometimes based on herbarium material (which often change with time); and the notes of different authors may be unreliable since they use different criteria for colour determination. Comparison of fresh flowers with colour charts is the preferred procedure to prevent such problems.

Several flowers have more than one colour (i.e. nectar guides) and furthermore some colours may change during the flower's life-span (see ref. 20). The colour-reflecting qualities of the flower should be evaluated in comparison with the natural background (15, 16, 23). Seasonal shifts of flower colours, and the occurrence and density of flowers of species with similar or different colours, should also be considered as a possible factor of interactions among plant species and pollinators.

The human visual spectrum (HVS) differs from insect visual spectrum (IVS) and together both constitute a good representation of the animal visual spectrum (AVS) (18). There is an overestimation of the importance of the UV reflection from flowers as a cue for bees, which should be treated as any other visual wavelength (16). Because bees can see as large an array of colours as humans, it is not enough to record the presence or the absence of UV. Full colorimetry range is needed for the understanding the significance of colours to insects (23, 24). The spectral reflectance cues should be used to represent the flower colour on a trichromaticity diagram (7, 18).

2.3 Colorimetric analysis of flower colours

This is an objective method in which the reflectance from the flower surface is measured using a spectrophotometer equipped with a special attachment for reflectance measurement (e.g. a Unicam SP 8000 recording spectrophotometer fitted with an SP 890 diffuse reflectance unit (25) or Bausch & Lomb Spectronic 20 reflecting spectrophotometer (26)). The results are expressed in a reflectance curve showing the percentage of total reflection in each wavelength sometimes including the UV regions.

In the case of multicoloured flowers, or if there are colour changes during the flower's life cycle, each colour area or change must be measured separately. When using coloured flower models in an experiment, it is advisable to measure the model's reflectance curve (in comparison to that of the living flower). Commercial paints which are similar to our eyes may have slightly different reflectance curves, especially in the UV area.

2.3.1 Flower photography using different filters, and the production of a spectral reflectance curve (5, 10, 23, 24)

Stages:

(a) preparation of a visible reflecting calibrated grey-scale

(b) photography or video-viewing of the flower alongside the grey-scale

(c) colorimetric analysis of the flower colour in the IVS (Insect Visual Spectrum) and the HVS (Human Visual Spectrum)

Protocol 1. Preparation and use of the grey-scale for measuring reflectance (24) and P. G. Kevan, pers. comm.)

Materials

- MgO and carbon powders
- liquid collodion (e.g. Fisher Scientific Co., Montreal C-407)
- propylene oxide
- filter paper (e.g. Balston, England #1)

Method

1. Mix the powders (MgO and C) by volume to get the necessary mixture (*Table 2*). The percentage of reflectance for each mixture is given in *Table 3*.

2. Dissolve 100 ml collodion in 100–150 ml propylene oxide.

3. Add 100 ml powder mixture, add mix to 100–150 ml. solvent and stir, till it is thin enough to paint.

4. Paint the filter paper and tape it down on a hard flat surface to prevent buckling and cracking during drying.

5. Cut small chips (1–2 cm) of all the mixtures, and stick them in ascending order on a strips of tape to get the full (1 to 14, *Table 24*) grey-scale.

The grey-scale is required to facilitate simple colorimetric analyses through determinations of reflectance in all wave bands across both the insect and human visual spectrum. Casper and La Pine (27) measured the UV reflection/absorption of the same flowers by three different methods and got different results. Caution should therefore be used in the interpretation of the UV measurement.

Table 2. Mixtures of MgO and carbon black used to make a grey-scale (from ref. 24, courtesy of P. G. Kevan)

Mixture No.	% by volume MgO	% by volume C
1	10	90
2	40	60
3	60	40
4	70	30
5	80	20
6	85	15
7	95	10
8	93	7
9	95	5
10	96	4
11	97	3
12	98	2
13	99	1
14	100	0

i. Camera

The camera (photographic or video) should be equipped with a quartz lens in order to capture a large portion of the UV spectrum, because most glass lenses only transmit long-wave UV (*c*. 320–400 nm), especially if coated for this purpose (although an ordinary lens has been successfully used (see, for example, refs 28–31)).

Some quartz lenses are chromatically adjusted for UV through the human visual spectrum (Zeiss UV Sonnar); others are not (Pentax Quartz Takumar); therefore, focus adjustments must be made for short wavelengths (23).

ii. Filters

A filter holder is attached to the front of the lens. It must be pieced together or built because such holders may not be available commercially (32). A series of broad-band monochromatic filters is used to capture the reflectance in each waveband. The following filters were used by Kevan (23).

Kodak 18A	ultraviolet	(glass)	300–400 nm
Kodak 35	violet	(gelatine)	320–470 nm
Kodak 98	blue	(gelatine)	390–500 nm
Kodak 65	blue-green	(gelatine)	440–570 nm
Kodak 61	deep green	(gelatine)	480–610 nm
Kodak 90	dark greenish amber	(gelatine	540–650 nm
Kodak 25	red	(gelatine)	580 nm
Kodak 70	dark red	(gelatine)	650 nm

[See Kodak (33) for more information]

Table 3. The percentage of light reflected at various wavelengths by the 14 chips of the grey-scale (from ref. 52, courtesy of P. G. Kevan)

Filter	Wave-length (nm)	% light reflection from 14 chips													
		1	2	3	4	5	6	7	8	9	10	11	12	13	14
18A	350	9.0	11.0	15.5	17.5	24.0	31.0	30.5	44.0	46.5	60.0	77.0	82.0	87.5	90.0
35	400	8.5	9.0	12.5	14.0	20.0	27.0	26.5	39.5	43.0	57.5	73.5	82.0	87.0	91.0
48	450	7.5	8.5	11.5	12.5	18.0	24.0	24.0	36.5	40.5	55.0	71.5	79.5	86.5	91.5
65	500	7.0	7.5	10.5	11.5	16.0	22.0	22.0	33.5	38.0	52.5	68.0	78.0	85.5	92.0
61	550	6.5	7.0	9.5	10.5	15.0	20.5	20.5	32.0	35.5	50.0	66.5	76.5	84.0	92.0
626	600	6.0	7.0	9.0	10.0	13.0	19.0	19.0	29.0	34.0	48.0	64.0	75.0	83.0	91.5
25	625	6.0	6.5	9.0	9.5	13.0	18.5	18.5	28.0	33.0	47.0	63.0	74.5	82.5	91.0
608	675	6.0	6.5	8.5	9.0	12.0	17.5	17.5	26.5	32.0	45.0	62.5	72.5	81.5	90.5

Note: Other filters have to be calibrated before use.

Other UV transmitting filters are the Schott UG 1 (32, 34, 35) and Kenko U 360 (transmitting long-wave UV 360 nm) (36). For visible light filter transmitting only, use a Wratten 2B filter (28). Sullivan (37) used the following filters transmitting red (Soligor 25AIR), yellow (Vivitar 8K2), blue (Vivitar 80C), UV blocking (Tibben), and visible light (Tiffen UV haze).

iii. The filter-factor calibration

Use an exposure bracket ranging from two times to eight times the exposure values without a filter to determine the best exposure of the grey-scale for each filter. Measure the light, for each filter, by using a through-the-lens light metre, to obtain the filter factor. Note that the filter factors will differ. In addition, at high altitudes, the UV and blue exposures are not as prolonged as at sea level. Video cameras have built-in exposure systems so that the filter factor is less of a problem, but tend to produce an over-exposure of dark images (5, 23).

iv. Focus adjustment

Use a 365-mm focusing attachment for the UV wave band and remove it later for photography (for Pentax lens). Make focus adjustments for the other shorter wavelengths (i.e. filters 35 and 48) by using the adjusted UV lens object distance and the normal lens object distance, and the results obtained by Kevan (23) for focus at those wave bands.

v. Film

The following films are used by various authors:

> Kodak Tri-X, forced to ASA 1600 rather than the ASA 400 designated (23, 31)
> Agfapan (ASA 100) (32, 35)
> Fuji SSS (ASA 200) (31)
> Kodak Plus-X (ASA 125) (29, 30)
> Fuji Neopan F (ASA 32) (31, 36)

vi. Photographing

(a) Fix the camera with the appropriate lens on the tripod.

(b) Focus the object through the view finder.

(c) Insert UV passing filters at front of lens.

(d) Do not forget to include a label with the object, as a record, that indicates which filter was used.

(e) Photograph the object without any filter and then progress through the filter series while adjusting the focus as needed (see *iv* above) for each of them.

(f) Take all the photographs (object and grey-scale) on a black background (black flock paper).

To avoid the use of a fixed camera or at least a tripod, which is often inconvenient for use in the field, Penny (32) used a modified version. The flower is brought into focus in the viewfinder and then the camera is held in position while the filter, also hand-held, is placed over the lens. UV exposure is then made, the filter is removed, and a paired full spectrum (visible light) exposure is made on the next frame. The difficulties of this technique are the small depth of focus that is obtained in close-up photography when wide apertures (ft. 4–4) that are needed for the UV exposure are used, and the impossibility of adjusting the focus after the filter has been placed over the camera lens.

vii. Illumination

The use of an artificial UV-rich light source (28, 29, 32), or even the use of natural direct sunlight may lead one astray through the appearance of artefacts and shadows. Error in interpreting putative UV reflectance may be produced by over-exposure without the use of a grey-scale to check exposure (16, 24).

viii. Preparation of the spectral reflectance curve

(a) Match the reflectance of the object (on the recorded image on the film or the video tape) with the reflectance of a chip of the grey-scale in the same frame.

(b) Convert the number of the matched chip (at the particular wave band) to the percentage of reflectance according to *Table 3*.

(c) Reconstruct the whole reflectance curve on the IVS and HVS.

(d) Use the reflectance data for the colorimetric analysis.

ix. Colorimetric analysis of flower colour (5)

The mixture of emitted or reflected wavelengths can be plotted according to the relative amounts of each of the primary colours. The proportions are then:

$$b = \frac{Rb}{Rb + Rg + Rr},$$

$$g = \frac{Rg}{Rb + Rg + Rr},$$

$$r = \frac{Rr}{Rb + Rg + Rr},$$

where R is the reflectance or (emittance) in *blue*, *green*, and *red*. A colour may be plotted by its coordinates (r, g) where r is the abscissa and g is the ordinate. The value of $b = 1 - (r + g)$. The coordinates can then be plotted on a triangular chromaticity diagram in which *b* has the coordinates (0, 0); *r* would be located at (1, 0); and *g* at (0, 1). White will be located at (0.333, 0.333), and yellow at (0.5, 0.5).

It is now possible to plot the chromaticity, or colour of the radiation, on a chart in which the primary colours are represented. This is done on a trichromaticity chart or triangle, of which the apexes represent the three primary colours for humans: blue (436 nm), green (546 nm), and red (700 nm), and for insects: blue (360 nm), green (440 nm), and red (588 nm).

By calculating the reflectance values for the three primary colours one may locate the flower reflectance on the colour triangle (IVS or HVS), whereby the different locations express the difference in colour of the same flower as seen by humans compared with what is seen by insects.

Menzel and his co-workers developed a flash-photometer (in the range of 300–700 nm, which is not available commercially) to measure the spectral emission of flowers. The colour loci of a flower are given in a physiological chromaticity diagram, in a colour hexagon as presented by Chittka *et al.* (38) and in a two-dimensional colour opponent diagram (39) for the honey-bee. The calculation of the loci uses as inputs the spectral sensitivity of the honey-bee photoreceptors (40); see review in ref. 7.

x. Developing
Silbergleid (41, 42) used Kodak Tri-X Panfilm which has been developed in an Acufine developer or Kodak Spectroscopic film developed in D-19 developer (see ref. 43 for various details).

2.4 Analysis of UV patterns in flowers

Protocol 2. Ultraviolet reflectance–absorbance photography (44)

Materials

- 35 mm single-lens reflex camera
- cable shutter release
- electronic flash
- tripod
- Kodak Wratten 18A filter
- UV-transmitting lens
- Kodak Tri-x Pan Film (ASA 400)
- pre-calibrated UV-reflecting grey-scale (see *Protocol 1*)

A. *Photography method*

1. Set up the tripod at an appropriate distance from the object.

2. Focus the image in the camera and photograph the specimen.

3. Mount the Kodak Wratten 18A filter without any other change, and photograph in the UV.

B. *Finding the suitable aperture*

1. Shoot an entire role of film at various f stops (shutter speed of 1/60 or 1/30 sec), with electronic flash as illumination source and the 18A filter in place.

2. Develop the film and print it in the usual manner. Choose the best quality photograph and indicate the appropriate *L* value.

3. Calculate the flash guide number (*F*) as follows (43) *F* = subject distance (*SD*) × lens aperture setting (*L*).

4. Use the *F* value to determine future exposure with the same camera, film, and flash attachment. When changing lens–subject distance, the new lens aperture is as follows: $L = (F)/(SD)$.

The wavelengths in the UV range being shorter than those of visible light, require a shorter focal length (i.e. shorter lens-film distance). The problem cannot be solved by refocusing with the filter attached, since the 18A filter is opaque. For cameras using an extension bellows on the stand, the focal distance can be corrected by decreasing the bellows extension, thereby shortening the focal length. Kevan (23) provides a graph from which the contraction of the bellows can be calculated without difficulty. When attempting close-up photographs, this problem is accentuated.

Protocol 3. Ultraviolet video-viewing (45, 46)

Materials

- portable video camera and video tape
- UV lens (e.g. Zeiss-Jena f4 60 mm UV lens with transmission wave of 330 to 400 nm)
- UV-transmitting filter (e.g. Schott UG 2; Corning C.S. 7–39; Kodak Wratten 18A).
- pre-calibrated UV-reflecting grey-scale (see *Protocol 1*).
- a 'black light' (e.g. Fluorescent General Electric BLE 1800B) for indoor viewing.

Method

1. Videotape the flower outdoors using either diffuse daylight from the sky or 'black light' when indoors.

Protocol 3. *Continued*

2. Place the grey-scale near the flowers while photographing for later comparison.

3. View the picture on the monitor. At this point you may photograph it in black and white to measure the relative UV brightness in comparison with the pre-calibrated grey-scale.

Note

If a UV-transmitting lens that is not colour-corrected for ultraviolet is used, then the visible and ultraviolet images will be separately discernible because they do not focus on the same plane.

Shmida and Menzel (47) suggest the following categories to classify UV patterns in flowers:

(a) Totally bright UV flowers
Flowers whose surface reflects UV illumination and thus look bright under this illumination.

(b) Dark UV flowers
Flowers which absorb UV illumination and look dark under this illumination.

(c) A combination of dark and bright UV; flowers which show a pattern of dark and light UV ('UV-bicolouration')

- contrast patches
- gradual patches
- veins and dark lines
- dark UV centre

(d) Mirror effect
Mirror effect can be detected by the UV lens with image intensifiers while rotating the flower.

These patterns are analysed in regard to the plant systematics, habitat, flower size and form, reward, visible pattern, pollinator, etc.

3. Flower shape and size

3.1 Introduction

A given flower has its unique physical characteristics such as size, shape, colour, spatial position, and odour (a blend of volatiles). While colours can be subjected to exact and objective photometric measurements (1, 5, 7, 48), and odours analysed with modern techniques (49, 50, 51) it is somewhat surprising that the study of flower size and form should have been a neglected topic. A summary of the common views on the importance of flower size and shape are presented in *Figures 1* and *2* and *Table 4*.

Frequently, the 'flower size' is expressed by the 'flower diameter' regardless of its form (13). Only a few studies have stressed the importance of the flower's planar projection (2, 53).

The form of the flower is termed as the 'figural quality' (57) to denote the ultimate shape of the flower (circle, ellipse, etc.). It has been found that honey-bees are very poor in distinguishing shapes (*Table 4*; but see ref. 53). The total contour length of the flower expresses the 'edginess' of the figure (57). It has been shown that butterflies and honey-bees (68, 69), as well as bumble-bees (12), prefer models with longer outlines.

The ratio between the area enclosed to the contour length expresses the 'figural intensity' (57). Anderson (58) proved that honey-bees compare shapes, remembering the contour density and not the contour length, as was formerly thought. For a general discussion it is quite satisfactory to denote the flower shape: bowl/dish, trumpet, flag cut, etc. (see ref. 12 and Chapter 1, *Table 2*).

For quantitative research, especially concerning the comparison of the magnitude of flower advertisement among species, there is a need for an exact and measurable parameter to express the flower's real size. A measurement of the flower's planar projection supplies a variable which exactly reflects the advertising surface of the flower (note that meanwhile, there is no useful way to evaluate the flower depth and its implications for the pollinator).

Many classical experimenters (see, for example, refs 54, 70, 71) used models (usually circles of different sizes) to examine various aspects of attraction and their relation to the target size. Currently, there are no known procedures that compare the relative advertisement of flowers of various sizes independent of their shapes or spatial position. Using the planar projection as a common criterion one can relate the advertisement area to the reward offered or by any other parameters (67).

3.2 Measurement of the flower's advertising area

The visitor arriving at a flower may perceive an image which is dependent upon the flower's tridimensional structure and the angle of approach. The same flower may have completely different sizes and shapes depending on various angles, or as a result of the flower parts' spatial position (72). To standardize measurements, the 'planar projection' of a flower is taken at an angle of 90 degrees to the flower's mouth plane. In this view, however, the depth effect (which is poorly understood) is neglected. One must also bear in mind that the visitor may approach the flower from various angles.

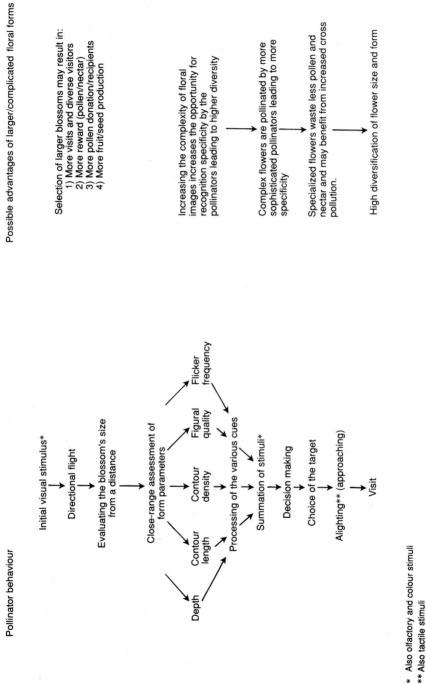

Figure 1. Insect reaction to non-colour visual stimuli of flowers.

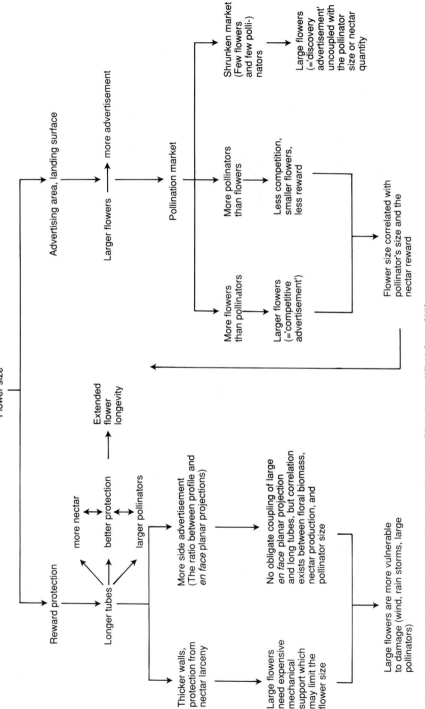

Figure 2. Implications for flower size in pollination. Slightly modified from (67).

Table 4. Parameters of flower size and shape and implications for pollinator behaviour

Category	Parameter	Definition	Implications/evidence
Flower size	Flower diameter	Edge-to-edge distance passing through the centre of the rounded shape	Visitation rate depends more on the target size (13, 52)
	Floral planar projection (A)	*En face* view from the pollinator's direction of approach = area of the photograph-like (eidetic) image (53)	Small flowers are usually organized in inflorescences → intensifying advertisement (12)
			Distance of attraction is directly proportional to the diameter (2)
			Diameter is a crude (but easy to measure) parameter for flower size becomes weaker along with the departure from a round shape
Flower form	Figural quality (T)	The ultimate shape of the flower (circle, ellipse, etc.)	Honey-bees are very poor in distinguishing shapes (1, 54, 55, 56)
	Total contour length (L) = figural intensity	The length of the form perimeter/edge	Expresses the 'figural intensity', e.g. the 'edginess' of the figure (57)
			Butterflies prefer models with longer outlines (68) and honey-bees have an innate and persistent preference for figures with longer outlines (54, 57)
			Bumble-bees are more attracted to more finely segmented flowers (12)
			Butterflies preferred figures with longer outlines (57)
			Bilateral/zygomorphic flowers have a longer contour length than actinomorphic ones. Bilateral flowers contain more information than similar reduced radial ones, with the same surface area (59)

Term	Definition	Notes
Outline irregularity	Contour length/enclosed area (L/A)	Bumble-bees prefer zygomorphic flowers over radically symmetrical ones, while honey-bees show reverse preferences (60)
		Irregular flowers have higher L/A ratios regardless of their form (54)
		Honey-bees compare shapes by remembering contour density and area (58)
= Contour density = Segmentation magnitude		Variations among species in the regularity of floral images are important in the specificity of recognition by the pollinators (61)
		Segmented blossoms may enhance the contrast with the background. Nocturnal insects react more strongly to such an effect resulting from strong blossom dissection (62)
		Honey-bees appear to prefer objects with irregular outlines over ones with regular outlines (63)
		Irregularities of an outline and its preference by bees presents an explanation for the evolution of the bizarre shapes in orchids (39)
Flower's profile planar projection (P)	The planar view of the flower's coloured area	
Side advertisement magnitude (SM)	The ratio between the en face projection (A) to the profile projection (P) (A/P)	High SM values are found in flowers pollinated by agents with a high manoeuvring ability (larger bees, hawk-moths, birds)
		The real advertising area which is exposed to foragers

Table 4. *Continued*

Category	Parameter	Definition	Implications/evidence
Flower depth	The distance from the flower mouth to its base		Bumble-bees prefer deep to flat models (63) as do honey-bees (2, 56, 64). 'The reason for this preference may be an instinctive or acquired knowledge that deep-lying sources of nectar are usually richer' (12)
Flower motion	Flickering and motion	The frequency of the image flashing as individual brightness changes (57)	Caused by movable parts (e.g. long lobes) and/or visual patterns (e.g. dots)
			A high frequency of flickering intensifies the stimulus as readily as the total shape does to our eyes; rapidly flying insects may resolve flicker frequencies as high as ten times more discriminating than the human eye (65)
			The attractiveness to flies of some Asclepiadaceae is apparently enhanced by their possession of vibratile organs (61) which also exist in some orchids (66)
			The widespread occurrence of flickering (pattern or real motion), plays an important role in the evolution of orchids (61)
			The flickering resolving ability increases the contrast (57), and hence the recognizability of flowers having this property (60)

3.2.1 The maximal attraction distance

For the purposes of measuring floral sizes with respect to bees, the following rule-of-thumb for recognizing their shortcomings is suggested (72):

$$\text{Visual acuity (resolving power)} \quad 1.4\,^\circ = \varrho$$
$$\text{Contrast sensitivity} \qquad\qquad 23\% = C_s$$

The latter value is highly conservative, but probably serves because most flowers are brilliantly coloured against a dull background, and overall contrasts are usually greater than 23% (47, 23, 16, 5). As more is learned about the visual system of insects, the values we have used can be modified to fit particular species and circumstances.

For example, if it is accepted that

(a) for reflections from an object to stimulate an individual ommatidium (facet) of an insect's eye, it must show a 23% difference from the background stimulating the adjacent ommatidia to be visible, and

(b) that the resolving power of the honey bee eye is 1.4 ° (101),

then a general formula which indicates the maximum attraction distance over which an object of a certain size can be detected by a single ommatidium (planar projected area normal to the surface of the ommatidium in question) can be given as follows:

$$D = \sqrt{A}/0.0186, \qquad\qquad [1]$$

where D is the maximum attraction distance and A is the planar projected area.

Figure 3 and the more general equation [2] provide the reasoning by which equation [1] is derived.

$$D = \sqrt{A}/C_s \cdot \pi \cdot (\tan g\ \varrho/2)^2, \qquad\qquad [2]$$

where D and A are as above, C_s is the contrasting sensitivity, and ϱ is the resolving power as represented by the angle of acceptance of a single ommatidium.

This approach is only useful for estimating the maximum distance over which an object, no matter how large or small, can be detected. The visualization of shape and pattern has not been addressed, but it follows that consideration of the number and pattern of ommatidia which are being stimulated simultaneously is required.

Pattern and shape recognition hinges on the concept of mosaic vision in insects (73). Thus, one can imagine that there is a minimum number of unicoloured, identically-sized, equilateral, hexagonal units (tiles) which must be employed to display a shape or pattern. Once that is known, the angle of acceptance that the arrangement of units (ommatidia) indicates can be used to calculate D.

109

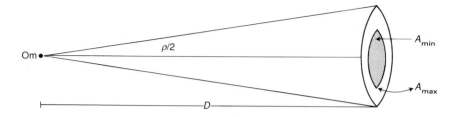

Figure 3. The object's size in relation to the field of vision of a single ommatidium. A_{min} = $A_{max}·\Delta l$ if $A_{min} < A_{max}·\Delta l$, if the object is below resolving power and too small to be distinguished.

$$D = \text{distance from the ommalidium to the object's centre}$$
$$Om = \text{ommatidium position}$$
$$A_{max} = \text{area of visual acceptance of ommatidium}$$
$$A_{min} = \text{area of object required to be visible}$$
$$\varrho = \text{acceptance angle of ommatidium (= resolving power of stationary eye)}$$

Using the above methodology and constraints, the maximum distance over which a blooming bush or tree of a known crown size attracts can be calculated, assuming that the approaching bee flies at the level of the crown.

Most bees fly at low altitudes, typically less than 10 m and most less than 4 m, above the ground. However, in calculating D, it is necessary to include flight height, because it influences the planar projected area of the patch impinging on the insects' ommatidium. The flower image area will change according to the tangent of the angle position that the insect makes with the patch (72).

Flower size and shape parameters are studied in relation to:

- type of pollinators ('the syndrome concept')
- magnitude of advertisement and pollinator's attraction
- reward structure (timing, quality, and quantity)
- diversity of visitors, pollinators, and thieves
- reproductive success and sexual selection

Protocol 4. Measurement of the flower planar projection

Materials

- a transparent hard board (glass or Perspex)
- plastic transparencies
- planimeter, or image analyser

Method

1. Place the flower on a table with its mouth surface facing up. Use a piece of plasticine to secure it. Fix a clear board 1–2 mm above the flower.

2. Put a plastic transparency on the board.

3. Draw the flower's *en face* contour from an angle of 90 degrees. Repeat the procedure with 10–20 flowers while keeping their exact spatial position even though they are not 'perfectly' symmetrical.

4. To get the 'profile planar project' of a deep (tubular) flower, place the flower in a horizontal position and repeat the drawing Be careful to match the two projections of each individual flower.

5. Use the drawings to measure the planar projection and the contour length (using a planimeter or image analyser). For the use and the implication of the various variables and their ratios see *Table 4*.

Flowers with long floral tubes are highly variable due to their spatial position; symmetrical actinomorphic flowers may have asymmetrical projections.

Other sources of floral size variability may be floral morphs (heterostyly, sexuality), age, and/or results of the physical environment (such as wind, rain, etc.). Data frequently collected from floras should be taken with care because there is frequently a tendency to draw 'an ideally symmetrical' flower which may be rare under natural conditions.

i. The implications of D—the maximal attraction distance

The calculated D (the maximal distance of attraction of a given flower's advertising area) provides a criterion (independent of shape and form) by which the hypothetical advertising efficiency of different flowers can be compared. The relation between D and other parameters of size and shape (e.g. contour length and the contour density) may be used to characterize pollination syndromes, to demonstrate a correlation between the flower shape and size, and to compare seasonal changes of advertising along ecological gradients, habitats, seasons, various floras, etc.

The measurement of the advertising area (and hence D) simultaneously with the calorific reward, and the frequency of the pollinator's visitation, may provide a ground basis for some basic evolutionary issues such as:

- the relation between flower size and the intensity of the advertisement expressed as the rate of visits
- the relations between the flower advertising surface area and the calorific reward

- a cost–benefit analysis of the advertisement magnitude vs. reward investment and visitation rate as a measure of fitness

ii. Side advertisement

Side advertisement is defined (67) as the case in which the profile planar projection of a given flower species is more or less equal or greater than the *en face* planar projection. A simple quantitative analysis (*Figure 4*) shows that side advertisement (when $A = P$) is maximal ($A + P = 1.41$) if the pollinator's approach is from an angle of 45 degrees. If the approach direction is between vertical (in an angle 90 degrees to the flower mouth), the advertisment is A, and if it is vertical to the profile, it is P. In any intermediate direction (unless the pollinator approaches from below whereby this situation needs further attention), the total advertisement is higher than A or P itself, and this is the real advantage of the side advertisement.

4. Scent as an attractant

Shape, size, colour, and fragrance all play a significant role in flower advertisement. Although many flowers are fragrant, only recently was a special scent gland discovered (74, 75). The great majority of scented plants studied show 'a diffuse' production of scents over the entire surface (74, 75).

Flower aroma is generally composed of a blend of several to many compounds (50) and each organ may have its own specific blend (49). Highly

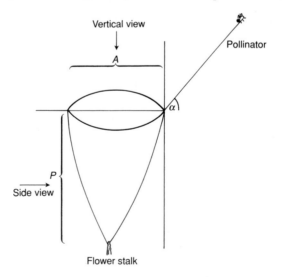

Figure 4. Side advertisement. α = the approach angle; A = the planar *en face* projection; P = the planar profile projection. If $A = P$ and $\alpha = 45\,°$, then A' = the relative area of A exposed to the approaching pollinator from angle α. $A' = A \sin \alpha$. If $\alpha = 45\,°$, $\sin \alpha = 0.702$, $P = A$, then A'' = the total advertisement of $A' - P'$ (P' = the relative side advertisement). $A'' = A' + P' = 2A \sin \alpha = 1.41\,A$.

volatile chemicals are regarded as long-range cues, but less volatile ones act at close range (76). Some chemicals are associated to 'scent guides' on the flower surface (74, 75) or may be emitted by pollen (Dobson, unpublished).

A scent organ which has a well-defined secretory tissue is called an osmophore (74, 75) and shows intense physiological activity. A specific case is the 650 or so orchid species found in the American tropics which are pollinated exclusively by Euglossine (Apidae) males (51). These males gather fragrant essential oils produced by special osmophores, especially on the labellum (74, 75). The exact reason why englossine males gather orchid fragrances has not yet been determined (51).

The flower-specific scent helps the pollinator to locate and recognize a particular flower, and the pollinator can so learn to associate the odour with the rewarding flowers. Specific aromas may enhance the fidelity of insects to particular flower species and they increase the efficiency of intraspecific pollen transfer (12). Robacker *et al.* (77) speculate that the floral aroma may serve as a dual olfactory communication system: (a) signalling the time when maximum nectar is available through maximum liberation of the specific odour component; but (b) giving misinformation by the scent emitted by non-rewarding older but receptive flowers which thus make use of the positive associative learning of their pollinators (honey-bees).

Deceptive odours are especially known in dung-flies' and dung-beetles' flowers (see 11, 78 for review), in which the flowers imitate the odour of the oviposition site of the female fly; and in sexual deception, in which flowers mimic a female insect odour. The best case studied is *Ophrys* (see refs 50, 79, and 80 for review).

Experiments/observations with scented models/extracts/plants parts are carried out for the following purposes:

(a) To examine pollinator preferences, discriminative ability, and/or behaviour pattern in response to odoriferous sources.

(b) To quantify/evaluate the significance/specificity/magnitude of the olfactory stimulus in relation to the other advertisement components *per se*.

(c) To compare the pollinator response to specific olfactory stimuli (isolated chemicals/blends for particular organ/flower/chemotype/morph/species).

(d) To understand the role of floral volatiles in pollinator attraction.

4.1 Field trials and bioassays using flower scents

The common methods for examination of the role of flower odours are as follows:

(a) Application of a specific chemical on a flower model or a piece of paper or cloth and observation visitor behaviour. By this method one can test the whole flower aroma, fractions of the blend, or isolated chemicals, and compare their relative attraction to pollinators (79, 81, 82).

(b) Manipulation of scents, colour, and rewards in artificial models.

(c) Choice experiments using different flower odours, odour fractions of isolated flowers or parts thereof, or chemicals (83).

(d) Associative learning experiments with various odours (84, 85).

(e) Field experiments in natural populations (86, 87).

Of all the flower cues, the olfactory stimulus is the most difficult to assess and identify. For behavioural experiments, there is a need to isolate the odour stimulus from other stimuli (colour, shape, and size). In the classical works of Knoll (88) and Kugler (2), simple devices were used to test olfactory attraction without visual stimulus: paper (non-transparent) bags were placed on plants while leaving holes for odour dispersion. To test colour without odour, the flowers were closed hermetically by a transparent cover to prevent odour emission. The experimenter exploited an existing learned experience of the pollinator, which associates odour with visual stimuli. These methods allow the possibility of assessing the relative importance of each stimulus. Other methods used to expose the pollinator to an unknown volatile and check its behaviour are shown in *Table 5*.

4.2 Olfactometry of flower scents

Pollinator reactions to floral scents could be evaluated by olfactometer (89, 90). The instrument consists (*Figure 5*) of two V-shaped tubes (G, G_1) which are separated by two cells (A, B). Fine nets (N, N_1, N_2) are inserted near the exits of these cells to prevent the insects from leaving the cells towards the air pump, and at the end of cell C to prevent the insects from leaving the system. A constant air-flow (e.g. by an aquarium air pump) is maintained via point E. The experiment is started by placing the flower (or flower part, extract, etc.)

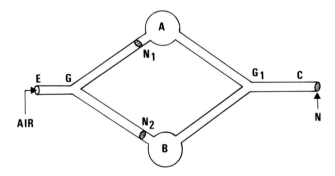

Figure 5. Glass olfactometer used to study the response of small insects to flower aroma. A, B = cells for holding the substrates under study; C = cells for the beetles; G, G_1 = glass tubes; N, N_1, N_2 = 40-mesh copper net; E = inlet for the air-flow. (Redrawn from ref. 90 with the publisher's permission.)

Table 5. Methods for testing flower odours in relation to pollination

Method	Advantages	Disadvantages	Selected references
Covering plants with perforated non-transparent material	Tests animal behaviour under natural and semi-natural conditions	Gives only a general idea about the scent's relative importance in the animal attraction	2, 91, 92
Preference experiments in cages	The animals can be observed individually under controlled experimental design	Behaviour in cages (frequently accompanied with artificial light conditions) may be different than under natural conditions	2, 69
Associative learning experiments	Gives a good control of the experimental design and the possibility to test comparatively various scent sources (chemical, natural blend, etc.)	So far, limited to honey-bees which can easily be conditioned. (Every experiment is limited to a few simultaneous odoriferous sources)	1, 84, 85
Attraction of pollinators to perfumed paper/models	The chemical/scent source may attract unknown visitors under natural conditions. Many different chemicals may be tested simultaneously. Easily combined with a test of other stimuli and very versatile for many manipulations, also under natural conditions	Gives only a general idea on the relative attraction magnitude of the examined material. The scented source has to be replenished frequently	2, 79, 82, 94
Manipulations with scent morphs	The odour factor can be tested almost separately from other cues under natural conditions in the same species. Some aspects such as relative attraction, foraging efficiency, and visitation patterns, may be related mainly to scent differences. Very versatile for many manipulations, also under natural conditions		93, 95
Olfactometry	Enable to test quantitatively the relative attractance of various odour sources	Suitable especially for small insects	90

in either cell (A or B), and then introducing the examined pollinator into cell C. The location of the insect is noted at proper intervals (such as 10–20 min) during the experiment period (1–3 h). If the insect is very small several of them can be tested simultaneously. The system has to be cleaned thoroughly (with chloroform and acetone) after every run to prevent contamination.

While running several experiments, temperature and illumination should be kept constant. By this method it is possible to measure the preference of the insects between two odour sources (e.g. two plant species, chemotypes, morphs, etc.).

4.3 Experiments with covered flowers

In this procedure (91) the plant/flower is covered by a non-transparent cover which hides the object but enables the odour to be emitted through suitable holes or openings. The insect reaction to scent emission is observed. Generally, a zigzag approach to the covered object (the zigzag approach is a specific response by an insect flying up in a vapour plume or concentration gradient of volatiles) indicates an olfactory attraction. The complementary experiment is covering of the object with an hermetic clear cover which prevents any odour transmission and checking for flower attraction without the presence of odour.

These two simple but basic experiments should supply the essential information for more further sophisticated analysis of fragrance and to indicate the relative importance of each cue. Both treatments (without odour and under non-transparent cover) should be compared to untreated flowers.

4.4 Localization of osmophores

Odour glands may appear on different floral parts. The simplest way to identify the odour source is to separate the flower into its parts, to accumulate the material from several flowers (if one flower is not large enough), store in a closed vial, and then to give it to several people to smell. Although this is very simplistic, it may yield some indications of the odour source. Staining flowers with neutral red (74, 75) will localize the odour secretion glands or areas. If the osmophores are borne on the flower's outer extremities (such as calyx ends or special appendages), it is easy to test the role of this gland in pollinator attraction. By simple manipulation, i.e. removal of the osmophores, it is possible to offer the pollinators intact flowers as well as treated ones, and then to compare the visitors' behaviour.

Protocol 5. Flower staining with neutral red

Materials

- 0.01 g neutral red powder
- 99 ml distilled water

Method

1. Dissolve the neutral red in the water. The dye is ready for instant use or can be stored at room temperature for further use.

2. Immerse the fresh intact flower samples in the neutral red dye for various durations (30 sec, 3 min, 1 h, 2 h, etc.). Take out of the dye and wash carefully with distilled water to remove excess dye.

3. Observe under a dissecting microscope. Osmophores are stained in a deep red colour.

Note

As neutral red stains positively for volatile oils the pollen-kit (surface pollen cement that forms a thick viscous coating layer over the grain) on pollen grains will also stain a deep brick red. Volatile oils are present in most plant tissues and neutral red will stain any region in which the epidermis has been cut, bruised or damaged. Therefore, it will stain the severed stalk of the flower or the base of any floral organ that has been cut and placed in the stain vial. Neutral red also stains nectaries, which should not be mistaken for osmophores. This method applies especially to pale-coloured flowers. If the background colour is dark, the contrast between the osmophores and the rest of the tissue is not readily discernible.

Protocol 6. Experiments on the associative learning of scents by honeybees (84, 85, 96)

Materials

The associative learning device (*Figure 6*) consists of a round polystyrene box (100 mm high, 230 mm diameter placed on a rotating table. Eight glass vials (25 × 60 mm) are pushed into and through the upper lid; when the box is closed, only the uncovered mouth of the vials are exposed. Each vial contains aromatic material and an inner, smaller vial (8 × 50 mm). A piece of perforated plastic net (e.g. Milar punch tape) is placed over the exposed mouth of the vials. This double-vial unit is called an 'aroma site'. The artificial feeder is an upside-down glass containing 70% sucrose in the centre of a 9 cm Petri dish covered with Whatman paper, No. 1.

Method

1. Recruitment: Put the artificial sucrose feeder on the feeding table, with a source of odour that the bees will associate with the sugar (e.g. excised flowers). Catch bees on the table, mark them as your 'experimental

Protocol 6. *Continued*

population'. Place the examined flowers (two different chemotypes, scent morphs or species, one of each of which was used while feeding) in random arrangement around the feeding table (2–5 m). Count the number of marked bees on the two examined types. Treat this as a choice experiment: known odour vs. unknown one.

2. Conditioning: Condition the bees one day before the experiment. During conditioning, put the odoriferous source (chemical essence, flowers, etc.) in the external vial in the device, and sucrose 50% in the inner vial. The associative learning device contains eight 'aroma sites', four with sucrose and an odoriferous stimulus, and four empty ones. Revolve the device frequently so as to ensure random location of each 'aroma site'.

3. Test: Replace the associative learning device which was used for the conditioning with one that contains four 'aroma sites' with only the odoriferous material, and four empty ones or with another odoriferous material. Revolve the device frequently. Each of the cruising bees which identifies a known odour reacts positively, lands on the punched net and extends its tongue. This response is considered a 'visit' (96). Change the punched net to prevent chemical marking by the bees' footprint substance (1), and possible resulting aggregation. Keep the nets and count them at the end of experiment (5–10 min).

Note

Capture and eliminate all insects of the 'experimental population' before the beginning of a new experiment. Leave a gap of at least three days and then establish a new experimental population. Stop the experiment for at least one day before changing the odoriferous material to which the bees are conditioned.

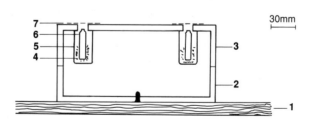

Figure 6. Cross-section of the associative conditioning experimental device (arena). **1.** = rotating table; **2.** = base of circular polystyrene box; **3** = polystyrene cover; **4** = aromatic material (excised flowers or inflorescences); **5** = external glass vial; **6** = internal glass vial; **7** = perforated cover screen. (84 with the publisher's permission.)

In this procedure you can compare the bee response and discriminative ability (chemically or systematically; e.g. morphs, congeneric species, leaves vs. flower of the same plant, etc.) between the two sources of odour that are closely situated. The principle is to condition the bee to one source of odour and test its ability to discriminate between this known material and a new one. The recruitment experiments are complementary, and serve as a kind of control under seminatural conditions.

Protocol 7 Attraction of pollinator to an odour source (79, 81, 92, 94)

Materials

Odour source (pure chemical, flower extract, or fresh flowers enclosed in perforated container), or filter paper pads (e.g. Whatman No. 1) or cloth (velvet).

Method

1. Saturate the filter paper with the odour source (if you squash a fresh flower, take care to use enough fresh material).
2. Expose the scented paper to a natural population of expecteed or known pollinators.
3. Replenish the odour source frequently (this depends on the evaporation period; e.g. 10 to 20 min).
4. Observe and score the various visitors and their behavioural patterns.
5. In studies of sexual attraction to *Ophrys*, a piece of black nylon velvet was used as a dummy to which the odoriferous compounds were added (see ref. 50 for a review).

Note

The evaporation function differs among blend or odours and chemicals may have different diffusion rates. This is difficult to take into account, but clearly may influence results. A parallel chromatographic analysis should be used before testing the comparative attraction of different odour sources. The specificity of odour attraction in flowers may be due to a certain blend of various compounds; thus, the attraction to a pure compound should be considered with caution.

4.5 Collection and analysis of flower volatiles (by H. E. M. Dobson)

Flower volatiles generally consist of a blend of diverse aliphatic, terpenoid, and aromatic compounds, but the representation of each varies extensively, both quantitatively and qualitatively, between species (see, for example, refs 51, 87). Among the various methods that have been used in the chemical studies of floral scents (see reviews in refs 50, 51, 87, and 97), the most accurate and reliable technique for collecting floral volatiles is that of headspace sorption, which is readily applied in both field and laboratory conditions. With this procedure, volatiles present in the air surrounding the flowers are trapped over a period of time on to an adsorbent material; they are then eluted from the adsorbent using a solvent. The resulting liquid sample is chemically analysed by gas chromatography (GC) for separation and tentative identification of compounds, and by GC coupled with mass spectrometry (GC-MS) for definitive identification. Details of the methodologies involved are discussed by Hills and Schutzman (98) and Dobson (87).

To collect volatiles, fresh flower material is placed within a chamber having two openings which are positioned such that air flows up through and over the flowers during its passage through the chamber. The chamber can be made of glass (e.g. 1- or 2-piece cylinders or flasks) or Perspex (e.g. square box), with the size depending on the kind of flower material to be sampled. Flowers can be sampled while still attached to the plant, which is least physiologically disruptive to the plant. In this case, either the whole plant is placed into the chamber or single flowers are sampled by covering them with open-ended glass cylinders. Alternatively, and of practical value in the field, flowers can be enclosed in bags through the bottom and exits through a hole made near the top. However, care must be taken to select a type of bag that neither releases nor adsorbs volatiles; some clear synthetic bags used for cooking food in the oven have been found to be satisfactory, but each brand must be tested beforehand. When working with whole plants is inconvenient, cut stems with flowers or inflorescences can be placed in a vase within the chamber, or in more extreme cases cut flowers or individual flower parts (including pollen) are put directly into small chambers. The method of choice depends on the plant species and on the chemical information desired. At all times, control samples should be collected simultaneously with the experimental ones from empty chambers and, if applicable, from green plant parts in order to determine the presence of contaminant non-floral volatiles. This is especially important when collecting volatiles in the field, where it is often difficult to pre-filter the air entering the chamber. During all steps of volatile collection and analysis, it is essential that contaminations be minimized by using clean equipment and preventing the samples or solvents from coming into contact with any material with oil base (e.g. cork, rubber, paraffin, some plastics); Teflon is a safe alternative.

Natural air is flowed through the chamber either by pressure (from an air cylinder) or preferably by suction (using a vacuum pump). To minimize the contamination of collected samples, incoming air is cleaned through a filter attached directly to the inlet chamber opening. Activated charcoal is effective and cheap for this purpose, and its low resistance to air-flow prevents changes in air-pressure within the chamber. Immediately at the chamber's outlet opening a cartridge is attached containing the adsorbent to collect plant volatiles. Approximate flow-rates for chambers 1–2 litres in size are 100–150 ml/min. The air-flow must be sufficient to maximize the amount of volatiles carried into the cartridge but low enough so that they are not carried through the cartridge and lost.

The duration of volatile collection that is required to obtain a sample strong enough for analysis depends on the plant species, since flowers vary widely in the quantities of volatiles emitted. For plants with weak to moderate emissions, sampling can be carried out over 12–24 h, but when the scents are strong 2–6 h may be sufficient. Plants also vary in the timing of volatile emission, and if one is interested in looking at possible periodicities in volatile production short sampling periods such as used by Matile and Altenburger (99) can be made in succession over 24 h.

Three main adsorbents are used for collecting flower volatiles: activated charcoal and the porous polymers Tenax GC and Porapak Q. Each has its own advantages and disadvantages, as well as particular procedures required for volatile desorption. Adsorbents are packed into narrow glass cartridges plugged with silanized glass wool on each side; in the case of charcoal, small thin filters (5 mg each) held inside a glass cartridge can be made or purchased. Sampling efficacy can be improved through the combination of different adsorbents, but cautionary steps may be called for (51, 98). Immediately after completing the volatile collection (during any waiting period the adsorbent cartridges must be kept cool), the volatiles are desorbed from the adsorbent, using either heat or more preferably solvents. The solvent samples can be stored at −20 °C. Depending on their strengths, the samples may need to be concentrated just prior to analysis.

Most criticial in the analytical procedures is the type of GC column used; columns of moderate polarity are best for the detection of a broad range of compounds, but each flower species has its own chemical characteristics. Likewise, a broad temperature programming (e.g. 50 °C initially, followed by an increase of 4–8 °C/min up to 250 °C) allows for the separation of volatiles having widely differing volatilities.

5. Experiments with flower models

Using artificial flowers enables one to change independently any components of the flower advertisement and/or reward. Flower advertisement is an integration of three main components: colour, shape, and odour. These elements should be considered along with their ecological background, such as the presence, density, and diversity of other flowering species, as well as pollinators, and the possible interaction among them. Experimental results using models should thus be applied to a natural situation with great care.

For practical reasons (availability, possibilities of control and manipulation, existing knowledge about the sensory abilities), most of the experiments are carried out with honey-bees. It is important to note the following precautions:

(a) The honey-bee is a social insect with an elaborate communication system; the behaviour of an individual is *not* totally discrete and independent. Furthermore, the previous experience of an individual has an important role in its behaviour.

(b) Solitary bees and bumble-bees may have different sensory acuities and abilities from the honey-bee, and there is no transfer of information between individuals.

(c) Differences in seasonality, habitat, body size, and interaction with other pollinators, may dictate or change the behaviour of bees on flowers. These may be expressed by nutritional needs, but also by heat and water balance, temporal and spatial distribution of sources, competition, etc. Broad knowledge of the specific insect's life-cycle is crucial in the interpretation of its behaviour, especially under artificial circumstances which may change its regular reactions (e.g. behaviour in narrow cages does not automatically reflect the natural situation).

(d) The same insect species may have different behaviour patterns in pollen-gathering, compared to nectar collecting, and these two activities may be carried out on different flowers.

(e) In an experimental design, one tries to isolate stimuli and their possible effects, but under natural circumstances all of the cues work simultaneously, and sometimes in a synergistic way. Thus, experimental results with models are not automatically applicable to the real and more complex natural behaviour.

(f) Naïve bees should be used in any experiments in which learning is involved.

(g) The insects should be reared so as to produce naïve individuals. Caught animals are subjected to unknown previous experiences and learning (a crucial point is some types of pollination by deceit and study of innate preferences).

(h) Honey-bees, as well as other bees (e.g. *Xylocopa* and *Bombus*), leave scent marks which transfer information to the next flower visitor.

(i) Honey-bees may change their response under artificial conditions (especially illumination and spectral composition, temperature, humidity).

(j) The flight height above the ground of males of some insects may affect its relative attractive ability regardless of any other cues (e.g. solitary bees which pollinated *Ophrys*; ref. 100).

(k) Under experimental conditions insects visit rather crude blossom models and mutilated blossoms apparently without discrimination as compared with the perfect specimens. It is questionable, however, whether this would apply under natural conditions with competition entering (ref. 12, p. 95).

(l) The reaction of a given insect to a certain model in the laboratory may differ from the reaction to the same model under field conditions.

References

1. Frisch, K. von (1967). *The Dance Language and Orientation of Bees*. Harvard University Press, Cambridge, Mass.
2. Kugler, H. (1943). *Ergebn. Biol.*, **19**, 143.
3. Harborne, J. B. (1982). *Introduction to Ecological Biochemistry* (2nd edn) Academic Press, New York.
4. Scogin, R. (1983). In *Handbook of Experimental Pollination Biology* (ed. C. E. Jones and R. J. Little), p. 160. Van Nostrand Reinhold, New York.
5. Kevan, P.G. (1983). In *Handbook of Experimental Pollination Biology* (ed. C. E. Jones and R. J. Little), p. 3. Van Nostrand Reinhold, New York.
6. Kevan, P. G. and Baker, H. G. (1984). In *Ecological Entomology* (ed. C. B. Huffaker and R. L. Rabb), p. 607. John Wiley, New York.
7. Menzel, R. (1990). *Color Vision in Flower Visiting Insects*. Forschungszentrum Jülich GMBH, Internationales Büro, Jülich.
8. Osche, G. (1983). *Ber. Dtsch. Bot. Ges.* **96**, 1.
9. Thorpe, R. W., Briggs, D. L., Estes, J. R., and Erickson, E. H. (1975). *Science (Wash.)*, **177**, 528.
10. Kevan, P.G. (1976). *Science (Wash.)*, **194**, 341.
11. Dafni, A. (1984). *Annu. Rev. Ecol. Syst.* **15**, 253.
12. Faegri, K. and van der Pijl, L. (1979). *The Principles of Pollination Ecology* (3rd edn). Pergamon, Oxford.
13. Waser, N. M. (1983). In *Pollination Biology* (ed. L. Real), p. 241. Academic Press, Orlando, Florida.
14. Willson, M. F. (1983). *Plant Reproductive Ecology*. Wiley Interscience, New York.
15. Menzel, R. and Backhaus, W. (1990). In *Vision and Visual Dysfunction* (ed. P. Gouras). Vol. 6, Chap. 15, Macmillan Press, Houndsville.
16. Kevan, P. G. (1979). *Am. J. Bot.*, **66**, 749.
17. Dafni, A., Bernhardt, P., Shmida, A., Ivri, Y., Greenbaum, S., O'Toole, Ch., and Losito, L. (1990). *Isr. J. Bot.*, **39**, 81.
18. Kevan, P. G. (1978). In *The Pollination of Flowers by Insects* (ed. A. J. Richards), p. 51. Academic Press, London.
19. Silbergleid, R. E. (1979). *Annu. Rev. Ecol. Syst.*, **10**, 373.
20. Gori, D. F. (1983). In *Handbook of Experimental Pollination Biology* (ed. C. E. Jones and J. R. Little), p. 31. Van Nostrand Reinhold, New York.
21. Kevan, P. G. and Lane, M. A. (1985). *Proc. Natl. Acad. Sci. USA*, **82**, 49.

22. Kornerup, A. and Wancher, J. H. (1961). *Methuen Handbook of Colour* (rev. D. Pavey). Eyre Methuen, London.
23. Kevan, P. G. (1972). *Can. J. Bot.*, **50**, 2289.
24. Kevan, P. G., Grainer, N. D., Mulligan, G. A., and Robertson, R. A. (1973). *Ecology*, **54**, 924.
25. Haslett, J. R. (1989). *Oecologia*, **78**, 433.
26. Macior, W. L. (1978). *Oikos*, **40**, 452.
27. Casper, B. B. and La Pine, T. R. (1984). *Evolution*, 38, 128.
28. Cruden, R. W. (1972). *Evolution*, **26**, 273.
29. Horowitz, A. and Cohen, Y. (1972). *Am. J. Bot.*, **59**, 706.
30. Abrahamson, W. and McCrea, K. O. (1977). *Rhodora*, **79**, 269.
31. Utech, H. F. and Kawano, S. (1975). *Bot. Mag. (Tokyo)*, **88**, 9.
32. Penny, J. H. J. (1983). *New Phytol.*, **95**, 707.
33. Kodak (1970). *Kodak Filters for Scientific and Technical Use*. Eastman Kodak Co.
34. Kugler, H. (1963). *Planta*, **59**, 269.
35. Giesen, Th. G. and van der Velder, G. (1983). *Aquat. Bot.*, **16**, 369.
36. Takahashi, H. (1984). *Bot. Mag. (Tokyo)*, **97**, 207.
37. Sullivan, J. R. (1984). *Am. J. Bot.*, **71**, 815.
38. Chittka, L., Beier, W., Hertel, H., Steinman, E., and Menzel, R. (1990). In *Brain–Perception–Recognition* (ed. N. Elsher and G. Roth), p. 195, *Proc. 18th. Göttingen Neurobiology Conf.* George Thieme Verlag, Stuttgart.
39. Backhaus, W. (1988). In *Sense Organs* (ed. N. Elsner and F.G. Barth), p. 219. George Thieme Verlag, Stuttgart.
40. Menzel, R., Ventura, D. F., Hertel, H., de Souza, J.M., and Greggers, V. (1986). *J. Comp. Physiol.*, **158**, 165.
41. Silbergleid, R. E. (1976). *Funct. Photog.*, **11**, 20.
42. Silbergleid, R. E. (1977). Unpublished mimeo.
43. Kodak (1972). *Kodak Data Book* M-27.
44. Hill, R. J. (1977). *Brittonia*, **29**, 382.
45. Eisner, T. M., Silbergleid, E. R., Aneshanley, D., Carrel, J. E., and Howland, H. C. (1969). *Science (Wash.)*, **166**, 1172.
46. Jones, C. E. and Buchman, S. L. (1974). *Anim. Behav.*, **22**, 481.
47. Shmida, A. and Menzel, R. (1991). Ecological aspects of U.V. reflection in flowers in relation to insect pollinators (MS.)
48. Daumer, K. (1958). *Z. Vergl. Physiol.*, **41**, 49.
49. Dobson, H. E. M., Bergström, G., and Groth, I. (1990). *Isr. J. Bot.*, **39**, 143.
50. Borg-Karlson, A. K. (1990). *Phytochemistry*, **29**, 1359.
51. Williams, N. H. (1983). In *Handbook of Experimental Pollination Biology* (ed. C. E. Jones and R. J. Little), p. 50. Van Nostrand Reinhold, New York.
52. Mulligan, G. A. and Kevan, P. G. (1973). *Canad. J. Bot.*, **51**, 1939.
53. Gould, J. L. (1985). *Science (Washington)*, **227**, 1492.
54. Hertz, M. (1935). *Naturwissenschaften*, **36**, 318.
55. Free, J. B. (1970). *Behaviour*, **23**, 269.
56. Bryan, A. D. (1957). *J. Anim. Ecol.*, **26**, 71.
57. Barth, F. G. (1985). *Insects and Flowers*. Princeton University Press, Princeton, NJ.
58. Anderson, A. M. (1977). *Anim. Behav.*, **25**, 62.

59. Davenport, D. and Kohanzadeh, Y. (1982). *J. Theor. Biol.*, **94**, 241.
60. Leppik, E. E. (1966). *Ann. Bot. Fenn.*, **3**, 299.
61. Davenport, D. and Lee, H. (1985). *J. Theor. Biol.* **114**, 199.
62. Vogel, S. (1954). *Blütenbiologische Typen als Elementeder Sippengliederung dargestalt anhand der Flora Südafrikas.* Fischer-Verlag, Jena.
63. Carthy, M. (1958). In Davenport and Kohanzadeh, ref. 59.
64. Hertz, M. (1931). *Z. Vergl. Physiol.*, **8**, 693.
65. Matthews, R. W. and Matthews, J. R. (1978). *Insect Behavior*, John Wiley, New York.
66. Proctor, M. and Yeo, P. (1973). *The Pollination of Flowers*, Collins, London.
67. Dafni, A. (1990). *Acta Hort.*, **28**, 340.
68. Ilse, D. (1928). *Z. Vergl. Physiol.*, **8**, 658.
69. Ilse, D. (1932). *Z. Vergl. Physiol.*, **17**, 537.
70. Hertz, M. (1934). *Bio. Zhl.*, **54**, 250.
71. Hertz, M. (1933). *Biol. Zhl.* **53**, 10.
72. Danfi, A. and Kevan, P. G. (1992). Advertisement in flowers in relation to floral, inflorescence and patch size, with respect to insect vision (MS.)
73. Mazokhin-Porshkyakov, G. A. (1969). *Insect Vision.* Plenum Press, New York.
74. Vogel, S. (1963). *Akad. Wiss. Lit. (Mainz) Abr. Math.-Naturwiss. Kl., Jahrgang*, **1962**, 599.
75. Vogel, S. (1990) *The Role of Scent Glands in Pollination.* Smithsonian Institution Libraries and The National Science Foundation, Washington, DC.
76. Bergström, G. (1978) In *Biological Aspects of Plant and Animal Coevolution* (ed. J. B. Harborne), p. 205. Academic Press, London.
77. Robacker, D., Meeuse, B. J. D., and Erickson, E. H. (1988). *Bioscience*, **38**, 389.
78. Meeuse, B. J. D. and Raskin, I. (1988). *Sex Plant Reprod.*, **1**, 3.
79. Kullenberg, B. (1961). *Zoologiska bidrag från Uppsala*, **34**, 1.
80. Borg-Karlson, A. K. and Tëngo, J. (1986). *J. Chem. Ecol.*, **12**, 1927.
81. Williams, N. H. and Dodson, C. H. (1972). *Evolution*, **26**, 84.
82. Pellmyr, O. Thien, L. B., and Bergström, G. (1990). *Pl. Syst. Evol.*, **173**, 143.
83. Dobson, H. E. M. (1987). *Oecologia*, **72**, 618.
84. Beker, R., Dafni, A., Eisikowitch, D., and Ravid, U. (1987). *Oecologia (Berl.)*, **79**, 446.
85. Pham-Delegue, M. H., Masson, C., Etievant, P., and Azar, M. (1986). *J. Chem. Ecol.*, **12**, 781.
86. Galen, C., Zimmer, K. A., and Newport, M. F. (1987). *Evolution*, **41**, 599.
87. Dobson, H. E. M. (1991). In *Modern Method of Plant Analysis* (ed. H. F. Linskens and J. F. Jackson), vol. 12, p. 231. Springer-Verlag, Heidelberg.
88. Knoll, F. (1926). *Abh. Zool.-Bot. Ges. Wien*, **12**, 378.
89. Peterson, A. (1953). *A Manual of Entomological Techniques.* Edward Brothers, Ann Arbor, Michigan.
90. Podoler, H., Galoon, I. and Gazit, S. (1984). *Acta Oecol. Oecol. Applic.*, **5**, 25.
91. Knoll, F. (1922). *Abh. Zool.-Bot. Ges. Wien*, **12**, 781.
92. Bino, R., Dafni, A., and Meeuse, A. D. J. (1982). *New Phytol.*, **90**, 315.
93. Galen, C. and Kevan, P. G. (1983). *Can. J. Zool.*, **61**, 1807.
94. Kullenberg, B. (1973). *Zoon., Suppl.*, **1**, 31.
95. Dobson, H. E. M. (1990). *Acta Hort.*, **288**, 313.

96. Waller, G. D., Loper, G. M., and Berdel, R. L. (1973). *Env. Entomol.*, **2**, 255.
97. Bergström, G., Applegren, M., Borg-Karlson, A. K., Groth, I., and Strömberg, S. (1980). *Chem. Scr.*, **16**, 180.
98. Hills, H. G. and Schutzman, B. (1990). *Phytochem. Bull.*, **22**, **9**.
99. Matile, P. and Altenburger, P. (1988). *Planta*, **174**, 242.
100. Paulus, H. F. and Gack, C. (1990). *Isr. J. Bot.*, **39**, 43.
101. Laughlin, S. B. and Horridge, G. A. (1979). *J. Comp. Physiol.*, **74**, 329.

5

Rewards in flowers

1. Introduction

While the various contrivances of advertisement serve as a means of attraction, the floral rewards have to supply some essential need of the consumer, to ensure the repeated visitation which can lead to pollination. Pollen and nectar are the main rewards offered by flowers to visiting animals in order to buy their services as pollinating agents (1). Nutritive rewards comprise both nectar and pollen as well as flower tissues as larval nutrition, food tissues (food scales, food bodies, non-fertile pollen, or pseudo-pollen), stigmatic fluid, and fatty oils. Non-nutritive rewards are nest materials (trichomes, resins, waxes, and corolla parts), shelter and warm resting places, sexual attractants and mating sites (see refs 1 and 2 for reviews, and *Table 1* for an overview).

2. Pollen as a reward

Pollen is a primary component of the flower reproductive system regardless of pollinator. Its packaging in small dispensable unit, which can be produced and exposed in controlled portions, in conjuction its with nutritive value, renders pollen as non-expensive currency to pay for pollination services, with minimal special adaptations. Although it seems that pollen is the most common and 'handy' reward exposed to a large array of pollinators, in reality rather few agents (mainly bees, flies, some butterflies, beetles, and bats) are adapted to digest pollen.

2.1 Pollen chemistry and energetics

When considering chemistry and/or energetics of pollen in relation to its value as a potential food-source for the pollinator, some caveats should be borne in mind.

(a) The use of bee-collected pollen is advantageous because large amounts can be obtained easily. But it should be remembered that the bees add significant amounts of sugar (as regurgitated nectar) to the pollen pellets.

Table 1. Types of reward in flowers and their implications in pollination

The reward	Chemical composition	Users	Ecological implications
Nectar	Carbohydrates, amino acids, lipids, antioxidants, alkaloids, protein, vitamins, and others	Almost all the known pollinator groups, especially for the forager's own consumption	The main calorific reward. Nectar volume, concentration, rate and rhythms of secretion are ± typical to the pollinator group. There is a correlation between the floral morphology (tube) and the presence/quantity/quality of the nectar
Pollen	Proteins, carbohydrates, amino acids, lipids, minerals, enzymes, pigments, and others	Especially insects, also used as a main food of the brood	In large grains starch is the energetic source but it is lipids in small ones. Pollen is a convenient food source which requires a minimum of adaptations on the part of the users; almost every mandibulate insect may use it
Stigmatic exudates	Lipids, sugars, amino acids, phenolics, alkaloids, and anti-oxidants	Insects	The main reward in only a few cases, especially in trap flowers
Floral tissues	Sugar, starch, protein, lipids, etc.	Insects (mainly beetles and bees). Bats	Special food-bodies, false anthers, or other unspecialized tissues (petals, etc.)
Oils	Saturated free fatty acids, diglycerides	Specialized female bees (Anthophorinae; Old Tropics, Rediviva; Mellittidae) South Africa	Only in specialized flowers mainly in the tropics. Produced in special structures (elaiophores). Oil collection requires the use of specialized structures formed by modified setae and fore tarsi
Perfume	Terpenes (e.g. cineole, geraniol, eugenol) Aminoid compounds (e.g. skatole), aromatic constituents (e.g. vanillin)	Male euglossine bees	Produced especially by Orchidaceae but also Araceae. The male stores the liquid in a special tissue in his hind tibia, but its exact metabolism and function is not known for sure

Resins and gums	Terpenes	Bees (Euglossini, Meliponini; Anthidiini); only females	Produced by a few plant species in the tropics. The bees use the material for nest construction
Brood sites		Highly specialized insects	Oviposition site and brood rearing in the flower (*Yucca, Ficus*): the adults are the most important or the only pollinators of the flowers involved
Shelter and heating	The flower offers shelter and/or heating	Insects (bees, flies, beetles)	Insects which sleep or stay at the flowers may get an energetic gain since the flower is warmer than the ambient temperature
	Parabolic flowers act as diaheliotropic solar furnaces	Insects	Insects bask in the flowers, which are warmer than the surrounding air, and make an energy gain
Meeting-places (rendezvous)	Higher chance for mates	Insects	Males which are seeking mates in the flower also pollinate them
Prey	The flower attracts prey	Insects	The predator serves also as a pollinator?

Sources: refs 1–10.

They probably cause changes in other constituents such as amino acids, vitamins, etc., due to their handling.

(b) Because pollen is hygroscopic, stored material may absorb a considerable amount of water and thus distort the recorded amounts of real water content of the pollen.

2.1.1 Pollen chemistry

From a chemical viewpoint pollen seems to be an excellent food source for flower-visiting animals because of its high nitrogen content and other essential chemicals (1). There is a correlation between pollen amount and its use as the only food source, especially in 'mess and soil' pollination of bowl-shaped flowers. Sophistication in flower structure and/or presentation of other kinds of rewards (especially nectar) is accompanied by a reduction in quantity of pollen (7).

For most pollination projects it is satisfactory to evaluate only the starch and oil content of the pollen. Starch content is visualized with IKI (see Appendix A2) and lipids with Sudan IV (see Chapter 3, Section 4.1). Pollen is generally designated as starchless or starch-rich, which usually also correlates to lipid-rich or lipid-poor, respectively (11).

Pollen that is sterile or non-viable for reproduction can still serve as a food source for anthophiles. Thus, a lack of stainability (see Chapter 3, *Protocol 1*) should not be automatically interpreted as useless for pollinators. Differences in fertility are expected in pollen flowers with differentiation of anthers to food pollen vs. reproduction pollen (7). Thus, it is essential to check both types of pollen separately for intine stainability, as well as for starch, oil, and presence of cytoplasm.

When considering pollen as a pollinator food source it should be borne in mind that pollen is primarily a microgametophyte which is part of the plant's sexual reproductive system and should function as such. Explicitly, pollen may contain various compounds (e.g. proline) which are essential to the pollen germination process and are not thought to result from selective pressure related to the pollinator's energetic and/or nutritional demands (8).

2.1.2 Pollen energetics

Starch and oil are the main calorific reserves of pollen. Baker and Baker (11) have shown that pollen rich in starch tends to have a low lipid content and is of lesser value as a food source for insects. Plant groups which offer pollen as the main or the only source of energy tend to have oil-rich pollen. In spite of the extensive data on the pollen oil–starch dichotomy, there is insufficient evidence to establish the selective forces which influence the presence or absence of starch in pollen (1, 12, 13, 14) in relation to mode of pollination (wind vs. insects) and the taxonomic plant groups. It should be noted that pollen has a higher energy content investment per gram of organic tissue than other plant parts (14).

Protocol 1. Measurement of pollen energetic value (by T. Petanidou)

Materials

- pieces of gauze or pollination bags
- Petri dishes, fine scissors, vials with cork caps
- sieves (mesh of 236, 132, 95, and 50 μm), acetone
- oven
- vacuum pump
- miniature bomb calorimeter (e.g. Gentry Instruments, improved Micro-bomb Calorimeter)

A. *Pollen collection*

1. Late afternoon before collection, in the field: Cover a sufficient number of flower buds of the plant species considered, using fine pieces of gauze or pollination bags, in order to prevent insect access and subsequent pollen contamination.
2. Following morning: Collect the anthers just before dehiscence in small Petri dishes using a very thin pair of scissors with fine tips.
3. Drying: Allow the anthers to dry in room temperature (open Petri dishes) but protected from dust (e.g. in cardboard boxes, or in a cupboard).
4. Pollen separation: When fully dehisced, remove the anther tissues from the exposed pollen by successive sieving. Use four different sizes of plastic sieves, with openings of 236, 132, 95, and 50 μm. During all sieving procedures be careful to avoid touching the anthers with the hands, but only with very thin and clean instruments (pair of scissors or forceps). Clean the instruments with acetone before use.
5. Purifying: Check the purity of pollen by examining the pollen under dissecting microscope, remove the contaminants with a thin pair of forceps.
6. Storage: Store the pure pollen in corked vials in a dry, fairly cool place (15 °C and 40% relative humidity) for not more than 40 days.

B. *Preparation of pollen tablets*

1. Removal of the water content: Place the pollen in an oven (35 °C for at least 2 days). Remove water by drying the pollen in a vacuum (at room temperature) for at least 12 h.
2. Tablet-making: Compress a certain amount of pollen in a tablet compressor and put, until combustion time, in an air-tight box containing silica gel for water absorption.

Protocol 1. *Continued*

3. Weigh the tablets just before putting in the microbomb. The weight should lie between 8.5 and 20 mg. Excess weight (>20 mg) is beyond the capabilities of the recorder.

4. Preparation of the microbomb for ignition: Join the two poles of the microbomb with platinum fuse-wire and then insert the tablet in a platinum plate in such a way that the upper free side of the tablet is in close touch with the wire. Close the microbomb tightly.

C. *Combustion*

1. Calibrate the instrument by burning (at least 10 runs) standard benzioc acid tablets (6318 cal/g). Repeat the calibration step every 30 pollen measurements (step 3).

2. Cool the microbomb (as taken from part B, step 4) sufficiently, so that its temperature baseline (taken when inserted in the calorimeter circuit) is linear and not abrupt.

3. Ignite of the pollen tablets (35 atm).

4. During ignition time connect the calorimeter to a special detector and record the temperature difference. The estimation of the heat released by pollen combustion is based on this graph.

5. Ash estimation: Weigh the platinum plate with the pollen combustion residue.

D. *Calculation*

Basically the energetic value is obtained by comparing the graph taken by pollen combustion with that of the standard (benzoic acid). Platinum correction in calories for residual fuse-wire = 1400 cal g^{-1}. (See the manufacturer's specific directions; ref. 15).

---

3. Nectar as a reward

Nectar is a sugary solution which is secreted by a special gland called a 'nectary'. The nectaries occur mainly in flowers but may occur on vegetative parts (16). Sugars are the major calorific components; sucrose, fructose, and glucose are the most common (4, 17, 18). For other nectar components and their possible significance in pollination see *Table 2*.

Deceptive, anemophilous, and hydrophilous flowers are nectarless. Nectar in trap flowers is regarded as providing substances for the pollinators which are imprisoned, rather than as a net energetic gain for them (23). This view needs testing by experiment. Nectarless flowers may appear regularly in nectariferous species and thus constitute a situation of automimicry. However, the evidence

Table 2. Nectar constituents and their ecological implications in pollination

Constituent	Range of concentration	Implications	Remarks
Sugars	5–75%	Main source of energy. Different pollinators prefer different ranges of concentrations	Sucrose, fructose and glucose are the most common, in ± fixed proportions for a given species. Sugar concentration affects the viscosity to which the pollinator is exposed
Amino acids	0.25–15.500 μmol/ml	The total amount reflects the pollinator type and the availability of amino acids in its diet	The complement of amino acids and their relative proportion are species-specific and heritable characteristics. Alanine and arginine are almost omnipresent
Proteins	Small amounts	A connection with enzymatic breakdown of sucrose (?)	
Lipids	Traces	Slight correlation with flowers pollinated by flies and bees, animals which digest lipids. A film of lipids on the surface may reduce nectar evaporation	Lipids in nectar are treated separately from oil glands which offer lipids as a separate reward
Antioxidants (e.g. ascorbic acid)	Traces	Protection of other nectar components (e.g. lipids) from oxidation	
Alkaloids	Traces	Deterring of non-pollinating visitors or thieves. Some moths and butterflies are deterred by alkaloids while bees and others might be less vulnerable	
Pigments	Low amounts	UV reflecting compounds render the nectar more visible to consumers (?)	

Table 2. *Continued*

Constituent	Range of concentration	Implications	Remarks
Vitamins	Traces	The adaptiveness of their presence is uncertain	Thiamin, riboflavin, nicotinic acid and others. Act as antioxidants
Essential oils	1–3%	Intrafloral olfactory guides?	The volatile blend of the nectar may differ from that of other parts of the plant
Polysaccharides	Low amounts	Contributing to the high viscosity or mucilage of 'vertebrate' flowers	

Sources: refs. 1, 2, 8, 9, 18–22.

is very limited (24), and needs further study. Heterogeneity in the spatial and temporal distribution of nectar volume and concentration has been widely studied, especially in relation to optimal foraging behaviour (25).

Because nectar is a product which is not a part of the plant's sexual system but a reward offered to a foraging agent, it is not surprising to find a tight correlation between the demands and the behaviour of the pollinator and the compositions, amount and rhythm of production of nectar. It is widely accepted that nectar plays an important role in plant–pollinator interactions reflecting co-evolution between the plants and their pollinators (3, 4, 18, 26, 27). The common nectar variables relevant to pollination are its concentration, volume, and sugar and amino acid content.

3.1 Nectar volume and concentration

3.1.1 Nectar volume and concentration measurements

Nectar volume and solute concentration are studied in relation to:

- the flower's life-span development, size, and position on the plant, and weather, soil conditions, and season
- limitation of seed production by pollination or other factors
- pollen flow, sexual selection, and the fitness of the male and the female functions
- the behaviour, and nutritional and energetic demands of pollinators

The method by which the nectar is extracted and measured is primarily dictated by the flower size, nectar volume, and solute concentration. A common method is to extract the nectar with calibrated micropipettes which work adequately for nectar volumes above 0.5 µl and concentrations below 70%. Routine concentration measurements are made with hand-held refractometers, with which small volumes of liquid (down to 0.1 µl) can sometimes be measured although the practical threshold is about 0.5 µl.

Extraction of nectar from small flowers needs special techniques (Section 3.1.2). The price paid for the measurement of nectar concentration, by pooling small volumes from several flowers, is the loss of information on the content of the individual flower. This method may distort real values, especially when different nectar volumes and concentrations from different flowers are involved due to their age, position, or previous exploitation by consumers.

Protocol 2. Measurement of nectar volume and concentration

Materials

- microcapillaries (1–25 µ)
- hand refractometer, modified for small quantities (0.1–0.5 µl)

Protocol 2. *Continued*

- ruler
- pH paper
- glucose test-paper
- micropsychrometer

Method

1. Extract the nectar with a suitable microcapillary (1–25 μl). If the nectar quantity is too small or its viscosity prevents free flow into the capillary (pipette), adjust the position of the pipette direction, gently squeeze the flower tube or apply suction with a flexible aspiratory tube attached to the pipette.

2. Measure the length of the nectar column (for volume calculations) if the microcapillary is calibrated.

3. In small flowers containing less than 0.1 μl (the minimal threshold for the refractometer, see below) gather nectar from several flowers (with awareness of the possible problems!).

4. Measure the nectar concentration with a pocket refractometer modified for small quantities (e.g. Bellingham & Stanley).

5. Afterwards, the nectar from the refractometer surface can be used for other examinations (e.g. pH and/or glucose test; *Figure 1*) especially if nectar quantity is small and if the extraction is tedious. (Be careful: exposed nectar changes its concentration rapidly, especially under high temperatures and low humidity.)

6. Calculate the nectar volume as follows

$$\frac{\text{length of the nectar column (mm)}}{\text{length of the pipette (mm)}} \times \begin{array}{c}\text{calibrated volume} \\ \text{of the pipette}\end{array}$$

$$= \text{nectar volume.}$$

7. Measure the temperature and the relative humidity (with a micro-psychrometer if possible) as close as possible to the flowers which were used, and simultaneously if possible.

3.1.2 Nectar extraction from small flowers

If the flowers are too small to be extracted by a regular microcapillary, the microcapillary tube may be gently drawn to a point over an alcohol lamp. Note that drawing out the microcapillary destroys its calibration, and nectar volume should then be assessed by the droplet technique (see *Protocol 3*).

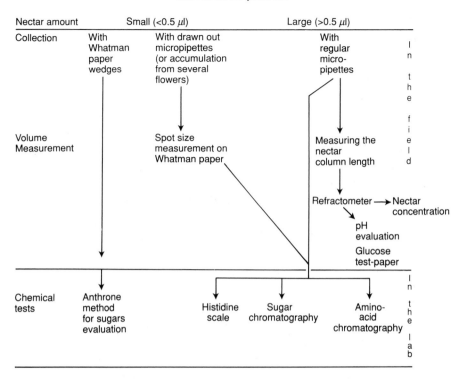

Figure 1. A schedule for nectar measurements and analysis.

Measuring a small quantity of sugar is also possible by washing a given number of flowers in a known volume of distilled water and then by chemical sugar analysis (28). Calibrated hypodermic syringes have been used to extract nectar (29). This method is suitable for tubular flowers and should be used with care so as not to damage the nectary tissue, especially if chemical analysis of the nectar is also involved. The nectar of small tubular flowers may also be extracted by centrifugation. A sample of 50 flowers should supply a measurable amount of nectar (30).

3.1.3 Use of the refractometer

- The optical glass surface of the instrument is very soft. Do not scratch it with the micropipette tip.
- The refractometer should be checked with a solution of known concentration every several hundreds measurements.
- If the nectar is too concentrated (and if only the low range 0–50% refractometer is available), the nectar may be diluted carefully with a known volume of distilled water, and mixed carefully.

137

- The higher the nectar concentration, the more viscous it becomes. This is accompanied by parallel volume reduction. Both factors may lead to reduced nectar harvest in the microcapillary and to less accurate measurements.

Note

- Insects can spit on concentrated or crystallized nectar and extract it even when it is beyond the reach of the microcapillary. Concentrated nectar above 75% can barely be extracted even with the help of an aspirator attached to the end of the microcapillary.

- The microcapillary may miss some nectar in a small flower. This can be avoided by drawing the tube out to a point, but with the loss of calibration. The amount of nectar that is reached by the microcapillary may also be influenced by the flower structure, such as the curvature of the flower tube.

- The measured concentration of solutes in the nectar depends on the concentration at which it is secreted and the extent to which evaporation or condensation concentrates of dilutes it after secretion (31, 32). Freshly secreted nectar (that is, new secretion after the flowers have been emptied) is needed to determine at what concentration the nectar is secreted under the observed circumstances (33).

- Other components such as amino acids contribute to the refractive index, affect the sugar estimation, and may reach values of 8 to 11% (34).

Protocol 3. Determination of volumes and sugar concentrations from nectar spots on paper (35)

Materials

- Whatman no. 1 chromatography paper
- methanol
- oxalic acid
- chloroform
- ethanol
- *p*-aminobenzoic acid (PABA)
- glacial acetic acid
- spectrophotometer

A. *Determination of nectar volume*

1. Spot the nectar on Whatman No. 1 chromatography paper.
2. Measure the spot diameter and compare it with *Table 3* to evaluate volume. This method is limited to spots up to 12 mm in diameter.

B. *Determination of sugar concentration*

(i) *Preparation of PABA staining solution*

Solution A: 75 mg oxalic acid dissolved in 15 ml ethanol

Solution B: 150 mg PABA dissolved in 25 ml chloroform and 2 ml glacial
acetic acid

ii. *Examination procedure*

1. Prepare Whatman No. 1 chromatography paper discs (about 6.4 mm
   diam.) using a standard paper punch.

2. Spot the sugar on to the discs and record its volume (the sugar spot need
   not fill the entire circle). Use a clean disc as a control.

3. Put each disc into a test-tube (e.g. 75 × 10 mm), cover with 50 µl of
   distilled water and allow to stand for 0.5–1 h.

4. Add 150 µl of PABA staining solution and mix well. Allow to dry. (Use a
   vacuum desiccator to speed up the process.)

5. When dry, heat for 10 min at 100 °C, until a brown colour develops.

6. Redissolve the stain in 1 ml methanol:water (1:1 v/v) and transfer eluted
   stain into semi-micro cuvettes. This stain is stable for at least 1 day. Read
   in spectrophotometer at 470 nm, using the control tube as the blank.

7. Sugar quantities can be obtained from suitable calibration curves. The
   ratio of sucrose to hexose in the standard curve should correspond
   approximately to that of the unknown.

8. Use three standards covering the sucrose:hexose ratios of >1, 0.5, and
   <0.5.

9. The procedure gives amounts of sugar in a known volume, i.e. weight/
   volume.

This method is accurate mainly in a sugar range of 18 to 68%; the
deviations from *Table 3* may be in the range of 28 to 58% (ref. 36 and *Figure
2*).

3.1.4 Nectar and the intrafloral microclimate

Since nectar is exposed to evaporation under the influence of environmental
variables (mainly temperature, humidity, wind speed, and solar irradiation;
Figure 3) it is essential to accompany every nectar measurement with
measurements of temperature and humidity (at least!) as close to the nectar
source as possible. Miniature devices for microclimate measurements such as
microthermocouples and portable psychrometers are available on the market.
Special instruments adapted for intrafloral microclimate measurements were
developed by D. Unwin and S. A. Corbet (39, 40).

Table 3. Correlation of nectar spot size with volume on Whatman No. 1 paper

Diameter (in mm)	Volume (μl)	Diameter (in mm)	Volume (μl)	Diameter (in mm)	Volume (μl)
0.1	0.01	4.1	0.650	8.1	3.3
0.2	0.02	4.2	0.700	8.2	3.4
0.3	0.03	4.3	0.750	8.3	3.5
0.4	0.04	4.4	0.800	8.4	3.6
0.5	0.05	4.5	0.850	8.5	3.7
0.6	0.06	4.6	0.900	8.6	3.8
0.7	0.07	4.7	0.950	8.7	3.9
0.8	0.08	4.8	1.000	8.8	4.0
0.9	0.09	4.9	1.050	8.9	4.1
1.0	0.10	5.0	1.100	9.0	4.2
1.1	0.11	5.1	1.160	9.1	4.31
1.2	0.12	5.2	1.220	9.2	4.42
1.3	0.13	5.3	1.280	9.3	4.53
1.4	0.14	5.4	1.340	9.4	4.64
1.5	0.15	5.5	1.400	9.5	4.75
1.6	0.16	5.6	1.460	9.6	4.86
1.7	0.17	5.7	1.520	9.7	4.97
1.8	0.18	5.8	1.580	9.8	5.08
1.9	0.19	5.9	1.640	9.9	5.19
2.0	0.20	6.0	1.700	10.0	5.30
2.1	0.215	6.1	1.765	10.1	5.45
2.2	0.230	6.2	1.830	10.2	5.60
2.3	0.245	6.3	1.895	10.3	5.75
2.4	0.260	6.4	1.960	10.4	5.90
2.5	0.275	6.5	2.025	10.5	6.05
2.6	0.290	6.6	2.090	10.6	6.20
2.7	0.305	6.7	2.155	10.7	6.35
2.8	0.320	6.8	2.220	10.8	6.50
2.9	0.335	6.9	2.285	10.9	6.65
3.0	0.350	7.0	2.350	11.0	6.80
3.1	0.375	7.1	2.435	11.1	6.99
3.2	0.400	7.2	2.520	11.2	7.18
3.3	0.425	7.3	2.605	11.3	7.37
3.4	0.450	7.4	2.690	11.4	7.56
3.5	0.475	7.5	2.775	11.5	7.75
3.6	0.500	7.6	2.860	11.6	7.94
3.7	0.525	7.7	2.945	11.7	8.13
3.8	0.550	7.8	3.030	11.8	8.32
3.9	0.575	7.9	3.115	11.9	8.51
4.0	0.600	8.0	3.200	12.0	8.70

Source: ref. 36, courtesy of H. G. Baker

Microclimate is relevant to pollination studies to determine:

- the chances of changes in nectar volume and concentration
- the patterns of daily and/or seasonal changes in nectar variables

- pollinator behaviour (time or activity, frequency, and duration of visits, foraging behaviour)
- nectar and pollinator energetics

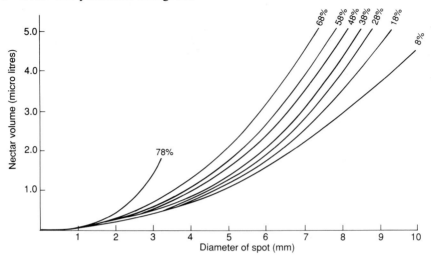

Figure 2. Nectar volume vs. spot diameter. (From ref. 36, courtesy of P. G. Kevan.)

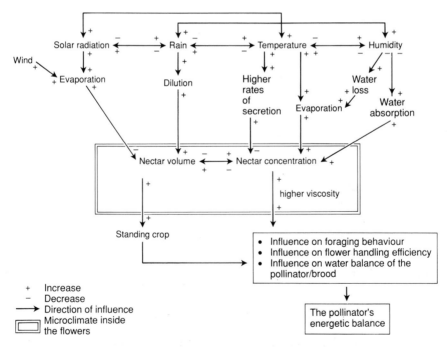

Figure 3. Weather variables and their influence on nectar volume and concentration in relation to pollination.

Protocol 4. Corbet's method for measurement of the intrafloral humidity (ref. 38, pp. 47–8; ref. 37, p. 37)

Materials

- thin (diam. less than 0.5 mm) epoxy-coated copper wire; entomological pins (size 0)
- potassium acetate 27%
- hand refractometer, modified for small amounts

Method

1. Insert a row (6–10) of entomological pins (size 00) 5 mm apart, on a piece of wood. Cut off the pins at a height of about 5 mm.
2. Bend the fine wire to create series loops about 5 mm apart.
3. Wind 10 cm of the wire about each pin to create a 'wiggly wire' containing several loops.
4. 'Fill' each loop with a small droplet of 1 µl potassium acetate solution with a microcapillary.
5. Fix the wire inside the flower, using a small piece of plasticine if necessary. Let it equilibrate with the ambient humidity.
6. Measure the intrafloral temperature, as close as possible to the measured droplets.
7. At the time of the measurement, 'harvest' a droplet from one loop with a microcapillary and measure its concentration in the refractometer.
8. Convert the potassium acetate concentration into the equivalent relative humidity value (*Table 4*).

If the flower is large enough, a very fine nylon net on a hydrophobic surface can be inserted horizontally into the flower. The potassium acetate droplets should be applied individually on to the surface. The potassium acetate droplet takes about 20–30 min to equilibrate with the ambient relative humidity. The droplet dries up completely at relative humidities below about 40%. It is suggested to prepare a number of replicas and then to sample them at intervals of 5 min to make sure that equilibrium has been reached. Potassium acetate is suitable for measuring humidities in the range of 40 to 95%. For high humidities (above 90%) the potassium acetate should be replaced by sucrose (30%; refs 37 and 38) for converting the reading from the refractometer to the value of relative humidity. This method gives a rough estimation of the relative humidity and, thus, should be used with great care.

Table 4. Conversion of potassium acetate concentration to relative humidity (courtesy of S. A. Corbet, pers. comm.)

%	RH	%	RH	%	RH	%	RH	%	RH
0	100	11	95.7	22	87.8	33	73.6	44	52.7
1	99.8	12	95.7	23	86.8	34	72.0	45	50.4
2	99.6	13	95.1	24	84.5	35	70.3	46	48.0
3	99.4	14	94.5	25	84.5	36	68.6	47	45.6
4	99.1	15	93.8	26	83.4	37	66.8	48	43.2
5	98.8	16	93.1	27	82.1	38	65.0	49	40.6
6	98.5	17	92.3	28	80.8	39	63.1		
7	98.1	18	91.5	29	79.5	40	63.4		
8	97.7	19	90.6	30	78.1	41	59.0		
9	97.3	20	89.7	31	76.7	42	57.0		
10	96.3	21	88.8	32	75.2	43	54.9		

% = potassium acetate reading in the refractometer
RH = percentage of the relative humidity

3.2 Nectar production rates and rhythms

Nectar production rates and rhythms (*Table 2*) are studied in relation to:

- pollinator activity patterns, frequency, efficiency, diversity, and rates of nectar consumption
- pollinator type (syndrome) and behaviour in relation to foraging energetics and reward structure

Protocol 5. Rates and rhythms of nectar secretion

Materials

- nets
- tags
- microcapillaries

Method

1. Cover the flowers or plants prior to anthesis and mark each flower individually.
2. Measure the nectar concentration and volume in a sample of 10–20 flowers each 2 to 3 h over the whole day. Use new flowers each time.
3. Parallel to steps 1 and 2, empty the the nectar and bag a sample of tagged flowers only for a limited time (30–120 min).
4. The measurement of an exposed flower at a given time will provide the nectar standing crop, while a parallel measurment of emptied and then bagged flowers will provide data on the nectar secretion rate at the observed period. (See also *Table 5*.)

Table 5. Nectar production components and their measurement

	Measurement procedure	Advantages/gained information	Disadvantages
Nectar volume	Extracting with calibrated micro-pipettes	Basic data for any further information on the nectar	Not accurate for viscous nectar or for small amounts (less than 1 µl)
Nectar concentration	Measuring with refractometer as sucrose equivalent	Value as a reward	The measurement reflects only sucrose equivalent, ignoring other constituents which may contribute to the solute's concentration level
Energetic value	The calorific value is calculated based on volume and concentration and the sucrose density	Essential information for any energetic studies	The calculations are based mainly on sucrose, although other sugars are present
Rate of consumption	Comparison of the standing crop of flowers exposed to pollinators vs. caged flowers at different times throughout the pollinator period activity time	Gives a real picture of nectar use under field conditions, usually accompanied with observation on pollinator activity, density, rate of visits or presence, and also micro-climate measurements	Change of microclimate under the cage may influence nectar production. Reabsorption if present is regarded as a consumption by insects; a measurement of the reabsorption is needed.
Secretion periodicity	Measuring of nectar volume at different times during the whole flower life-span	Generally, there is a correlation with the pollinators' activity	Nectar reabsorption may mask/change secretion in absence of consumers, especially if there is more than one daily peak in secretion. Should be studied with care to distinguish between fluctuations due to internal and external conditions

Concentration of freshly secreted nectar	Flowers are emptied with micro-capillary or absorbent paper. After a given period the freshly secreted nectar is collected again and then measured	Knowledge about fresh nectar permits the evaluation of changes caused by external factors	Evaporation or condensation after secretion depends on temperature and humidity in the immediate environment
Rate of secretion/production	(a) Removal of nectar at known intervals and measuring of the new production (b) Harvesting the nectar of chosen different caged flowers at various time intervals	Gives the rates of production under the removal regime Samples at different times may mask the changes caused by microclimate variables	Removal of nectar may increase or decrease the subsequent secretion Repeated nectar extraction may mask nectar reabsorption and thus give higher rates for secretion
Standing crop	Calculation of mg sugar per flower at a given time	Standing crop depends on the rate of secretion in relation to use. Gives an instant still picture on the level of exploitation of the available nectar by pollinators. Measuring this value at different pollinator density/activity in relation to micro-meteorological variables at different times may evaluate the production/consumption rate related to pollinator activity by comparison with unvisited flowers	Valid only for the moment of the measurement

Possible sources of intraspecific variability in nectar production can be caused by:

(a) Differences between flowers on the same plant due to position on the flowering stem and/or exposure to ambient microclimate (especially in large canopies).

(b) Differences between individuals and populations of the same species ('cold' and 'hot' plants).

(c) Daily and seasonal variations:
- different flowers on the same plant may start secretion at different times
- day-to-day variation in weather may cause shifts in the pattern of nectar characteristics (*Figure 3*)

(d) Morphological and phenological characteristics:
- large flowers may secrete more nectar than smaller ones at intra- and interspecific levels
- the flower phenological stage (e.g. protandry, protogyny, ageing) influences nectar production

(e) Experimental manipulations:
- Nectar extraction may enhance or reduce subsequent secretion rates. Reabsorption of nectar may elevate the apparent secretion rate in repeated nectar removals.
- Potted plants (in experimental plots, greenhouses, or growth chambers may differ in nectar production from plants in natural conditions.
- Cut flowers generally produce less nectar than flowers on the parent plant.
- Microclimate around or inside the flowers of covered plants influences nectar secretion and nectar characteristics.

(Main sources: refs 25, 31, 39, 40, 41).

To overcome these possible sources of variability the following steps are recommended:

- Measure the microclimate around or within the flowers simultaneously in covered and uncovered plants. The temperature and/or humidity is usually different inside nets or bags, etc. (40).

- Compare all-day covered flowers to short-term covered flowers at different times during the day. This will show the rate of change in the quantity of nectar, and therefore the net rate of secretion and/or resorption, in all-day covered plants (33).

- The nectar measurment should cover the flower's entire life-span.

- Generally, an additional series of nectar measurement is carried out in flowers freely exposed to visitation by nectarivores. The comparison between nectar production under nets or bags and in freely visited flowers may be used to indicate actual nectar exploitation in relation to animal activity.

- A parallel census of the flower visitors, their frequency and diversity (see Chapter 6, *Table 2*), during caging of flowers at different times, may provide information on nectar consumption of animals active at various periods of the day (day- vs. night-, or morning- vs. noon- or afternoon-visiting animals).

- Since nectar production rate is not always constant, it should be estimated for the whole day(s) and not only for short intervals.

3.3 Nectar energetics

3.3.1 Energetic value of nectar

The readings in most refractometers are in sucrose equivalents expressed as milligrams of sugar per 100 mg of solution. They can be converted to milligrams of sugar per flower by converting the measured sucrose equivalent to g/litre and multiplying this value by the nectar volume (42). The conversion tables of sucrose concentrations to density are provided in *Table 6*. The equation for the calculation of the sugar amount is:

$$\begin{array}{l}\text{Milligrams} \\ \text{in } A \\ \text{volume} \\ \text{of nectar}\end{array} = \dfrac{\begin{array}{l}\text{\% sugar-reading in} \\ \text{the refractometer}\end{array}}{100} \times \begin{array}{l}A \text{ volume} \\ \text{in } \mu l\end{array} \times \begin{array}{l}\text{Density of} \\ \text{sucrose at} \\ \text{the observed} \\ \text{concentration.}\end{array}$$

For energetics calculations:

$$1 \text{ mg sugar (sucrose)} = 4 \text{ cal} = 16.8 \text{ joule.}$$

Cruden and Hermann (40) developed a regression equation relating the sugar quantity (y) to the measured concentration (x) ($y = 0.0046x + 0.9946$). Their equation is based on measuring the concentrations of known dilution series of sugar solution.

Prŷs-Jones and Corbet (ref. 44, p. 74) present the following equation to calculate the sugar quantity: $\hat{P} = 0.003729/C + 0.0000178C^2 + 0.9988603$, where $\hat{P}$ = the estimate of density for a given value of C (*Table 6*) and C the weight of the sucrose per 100 g solution (= the refractometer reading).

Note

- Amino acids (e.g. proline) also have caloric values and could be used directly as a source of energy (34, 40). Lipids are also used immediately as a rich energy source (40) but are normally presented in very small amounts in nectar.

Table 6. Density of sucrose solution at 20 °C at various concentration

% sucrose by wt	Density	% sucrose by wt	Density
0.50	1.0019	24.00	1.1009
1.00	1.0039	26.00	1.1102
1.50	1.0058	28.00	1.1195
2.00	1.0078	30.00	1.1290
2.50	1.0097	32.00	1.1386
3.00	1.0117	34.00	1.1484
3.50	1.0137	36.00	1.1583
4.00	1.0156	38.00	1.1683
4.50	1.0176	40.00	1.1785
5.00	1.0196	42.00	1.1889
5.50	1.0216	44.00	1.1994
6.00	1.0236	46.00	1.2100
6.50	1.0257	48.00	1.2208
7.00	1.0277	50.00	1.2317
7.50	1.0297	52.00	1.2428
8.00	1.0317	54.00	1.2540
8.50	1.0338	56.00	1.2654
9.00	1.0358	58.00	1.2770
9.50	1.0379	60.00	1.2887
10.00	1.0400	62.00	1.3006
11.00	1.0441	64.00	1.3126
12.00	1.0483	66.00	1.3248
13.00	1.0525	68.00	1.3371
14.00	1.0568	70.00	1.3496
15.00	1.0610	72.00	1.3623
16.00	1.0653	74.00	1.3751
17.00	1.0697	76.00	1.3880
18.00	1.0741	78.00	1.4011
19.00	1.0785	80.00	1.4142
20.00	1.0829	82.00	1.4275
22.00	1.0918	84.00	1.4409

Source: ref. 43, p. d–320 (with the publisher's permission.)

3.4 Nectar constituents

Nectar constituents are studied in relation to

- pollination syndromes
- pollinator energetics and nutritional demands
- plant phylogeny

It is advisable to analyse freshly harvested nectar at the peak of its secretion, or to elute it from a filter paper on which it was spotted previously but immediately after extraction from the flowers and then dried. If the nectar stands in a liquid condition, expecially at high ambient temperatures, it may undergo changes such as: breakdown of the sucrose into glucose and fructose or changes in the amino acid content (by microorganism activity and changes of pH).

3.4.1 Sugars

Protocol 6. Determination of the total carbohydrate content of small amounts of nectar (ref. 45; ref. 46, p. 261)

Materials

- Whatman No. 1 filter paper; block of plastic foam
- fine insect pins (No. 0)
- pointed forceps
- anthrone (9 oxyanthracene)
- concentrated H_2SO_4 (95%)
- vortex
- spectrophotometer

Preparation of the anthrone reagent

1. Carefully prepare H_2SO_4 95% by addition of 100 ml of the concentrated acid to 5 ml of distilled water while cooling.
2. Dissolve 2.0 g of anthrone in 1 litre of 95% H_2SO_4 (prepare fresh).
3. The reagent darkens with time, so standards must always be run.

A. *Nectar extraction*

1. Cut a Whatman No. 1 filter paper into small wicks (2 × 8 mm) tapering at the end.
2. Prepare in advance a stock of wicks impaled on fine insect pins.
3. Insert the pinned wick into the flower and absorb the nectar (use forceps to avoid touching the wick with fingers). Replace the wick if fully saturated.
4. Prepare the outline of the sampling scheme in advance on a piece of ruled paper clipped to a block of plastic foam. After each sample, pin the wick in its appropriate place on the sheet and allow it to dry in the air for later analysis at the laboratory.

B. *Colorimetric analysis*

1. Dissolve the sugar from the air-dried wicks by agitating the wicks in 5 ml of boiling distilled water for 1 min.
2. Place 2-ml samples of the resulting solution, reagent blanks, and a series of sugar standards (see below), in a screw-cap test-tube in an ice bath.
3. Add 4 ml of fresh anthrone reagent, cap, and agitate the tubes, and place them in a boiling-water bath for 10 min. Let them cool to room temperature, then read absorbencies in a spectrophotometer (at 620 nm).

Protocol 6. *Continued*

C. *Calibration of the sugar standards*

1. Prepare the range of fresh sugar standards ranging from 10 to 300 mg of total sugar per 2 ml of distilled water with equal amounts of fructose and glucose.

2. If the nectar is too concentrated, dilute it to bring it to the range of the absorbency readings of the standard series.

3. Compare the results of absorbency readings of the sample to the standard series and estimate the amounts of sugars.

Note

The anthrone reacts with all carbohydrates to give a characteristic blue colour. It reacts with mono-, di-, and polysaccharides, but the colour produced is not the same for different carbohydrates. A previous knowledge of the nectar sugar composition is needed for preparation and calibration of the standards.

Protocol 7. Paper chromatography of nectar sugars (3)

Materials

- Whatman No. 1 chromatography paper filter; fluorometer (e.g. Turner, Model III); vacuum desiccator
- ethyl acetate; *n*-propanol; methanol 50%
- 75 mg oxalic acid; 15 ml ethanol 95%; 150 mg *p*-aminobenzoic acid; 25 ml chloroform; 2 ml acetic acid

Preparation of sugar staining reagent (47)

Solution A: 75 mg oxalic acid dissolved in 15 ml ethanol
Solution B: 150 mg *p*-aminobenzoic acid in 25 ml chloroform and 2 ml acetic
              acid.
Mix solutions A and B just prior to use.

Chromatography procedure

1. Spot the nectar on Whatman No. 1 chromatography paper.

2. Cut out the spot and dilute the nectar in a test-tube with 1 ml of distilled water for 24 h at room temperature. Take the piece of paper out and concentrate the solution with a vacuum desiccator.

3. Re-spot the nectar on Whatman No. 1 chromatography paper.

4. The solvent solution is *n*-propanol:ethyl acetate:water (14:4:2 v/v/v); run for 65–72 h.

5. Run a standard mixture of known sugars on each paper as a control.

6. Air-dry the paper with a hair dryer (cold air!) and dip the paper in the staining reagent.

7. Air-dry the paper again, then heat it at 110 °C for 20 min, by which time pentoses become red and all other sugars produce a brown colour.

8. Outline the spots under UV illumination in which all sugars fluoresce clearly.

9. Cut the outlined spots individually, dilute into 4 ml of 50% methanol, and measure the fluorescence in a filter fluorometer (e.g. Turner model III with primary filter Corning 7–60 and secondary filter Kodak 12).

10. Prepare a calibration curve by running reference sugars of known concentration using the same chromatographical procedure.

Note
- As the relationship between sugar concentration and fluorescence value is linear only in diluted solutions, it is necessary to dilute the stained sugar with 50% methanol.
- The methods are sensitive to sugar concentrations of 0.5 mg/ml and upwards.

Protocol 8. Thin-layer chromatography (TLC) of nectar sugars (S. A. Corbet, pers. comm.; ref. 48, p. 164, modified)

Materials

- silica gel TLC plates (e.g. SIL G/25/UV, CAMLAB, No. MN–809–024)
- CAMLAB pocket chromatography chamber
- sprayer or perfume atomizer
- chloroform, methanol (analytical), acetone
- sugar references (sucrose, fructose, glucose, maltose, raffinose, etc.)
- aniline–diphenylamine
- orthophosphoric acid (85%)
- Whatman No. 1 paper (50 × 157 mm) sheets

Preparation of the sugar colouring reagent (or prepared spray, e.g. Sigma A 8142):

1. Dissolve 4 g aniline-diphenylamine, 4 ml aniline, and 20 ml *o*-phosphoric acid in 200 ml acetone.

2. Heat for 10 min at 45 °C in water bath

Protocol 8. *Continued*

Chromatography procedure

1. Mark with a fine pencil a line 1 cm from each side of the silica gel TLC plate. Mark five equidistant spotting points along the baseline.

2. Spot a tiny droplet (about 0.1 µl with a micropipette or fine paint-brush) of nectar and each of the reference solutions on the marks on the baseline. Let each droplet dry then reapply 3–4 times, but keep the spots as small as possible.

3. Insert the silica gel plate into the chamber while a Whatman No. 1 paper of the same size is beneath the plate. One end of the paper is attached to the back of the plate at its distal edge and the free end is laid at the proximate liquid chamber. Close the chamber tightly.

4. Add 5 ml of the running solvent (40:60 chloroform:methanol; v/v) through a hole in the upper cover of the chamber, using a syringe, and cover the hole with tape to prevent evaporation.

5. Be careful not to wet the plate while filling the liquid cavity. Let the inner atmosphere be saturated for 5 min and then put the chamber in a vertical position.

6. Let the front proceed (25 to 40 min) until it is 2 cm from the upper end of the plate. 8–10 cm of running is also sufficient. Mark the front line.

7. Uncover and dry the plate with cold air from a hair dryer.

8. Spray the plate gently but uniformly with the aniline–diphenylamine (a small perfume atomizer can be very useful for this). Dry the plate again for several minutes in a fume cupboard or well-aerated space.

9. Heat the plate at 85 °C for 10 min. The different sugars will appear as various colours and rates of flow. Compare the nectar to the references carefully, especially for the blue spots.

10. Photocopy the plate and mark the results. The original colours and the whole plate will turn blue after several weeks.

Notes

(a) The running solvent has to be freshly prepared every time. The ratio between the chloroform and methanol can be changed to ratios from 50:50 to 70:30 (chloroform:methanol).

(b) Use only analytical samples of saturated solutions for the sugar reference solutions.

(c) Use a small droplet of nectar (0.1 µl) and uniform droplets of reference solutions. A very fine paint-brush with few hairs is very efficient for this purpose. Wash and dry the brush after every sample.

(d) Measure the sugar concentration of the nectar and the reference solutions. Try to work at as close a range of concentration of the nectar as well as of the references as possible.

(e) Changing the ratio between the chloroform and the methanol will change the relative rates of flow (RF) among the sugars. For example 60:40 methanol:chloroform will sharpen the RF differences between the sucrose and the fructose.

(f) Use only freshly harvested nectar. The monosaccharides may increase in nectar that has been allowed to stand as a liquid because of the breakdown of disaccharides. You may overcome this problem by collecting the nectar *in situ*, sealing the capillary (even with plasticine), and transporting it to the laboratory on dry ice. Freeze it until the analysis. Another possibility is to put it on a filter paper, dry it, and later dilute it in water in the laboratory immediately prior to examination.

(g) Microorganisms may change the sugar and amino acids composition of nectar.

This qualitative procedure can be done with simple means, even under field conditions. If pocket chambers are not available, they can be replaced by a beaker large enough to contain the plates. For other, more complicated quantitative analysis of sugars, other chromatographical methods can be used such as HPLC (see, for example, ref. 49), or GC (see, or example, ref. 50).

Table 7 summarizes the advantages and the disadvantages of the described methods for the determination of the calorific value and sugar content of nectar.

Comparison data for sugar chromatography:

Sugar	Colour with AD	Rate of flow
Fructose	Pink	~ 70
Melibiose	Blue	~ 60
Glucose	Dark blue	~ 70
Maltose	Blue	~ 50–60
Sucrose	Black	~ 50–60
Raffinose	Blackish-blue	~ 20–30

i. Glucose evaluation of nectar

Commercial glucose test paper is used for detection of sugar in urine as a test for diabetes. This examination gives an evaluation of the glucose content in the range of 40 to 250 mg glucose per 100 ml solution. It is useful for quick determination of glucose-rich nectars and for indirect evidence for sucrose degradation by microorganisms under natural conditions by comparison of fresh and old secreted nectars.

Table 7. Methods for determination of nectar concentration and sugar content

Method	Advantages	Disadvantages
The 'wedge wick' method (45)	Suitable for small amount of nectar. Handy and easily carried out in the field. Nectar samples are free from contamination with pollen or stigmatic exudate (if any). Less residual matter left over	Less information on volume and concentration. Previous information on the nectar components is needed for calibration of the standards
Micropipettes and refractometer	Instant information on volume and concentration. The nectar may be used for further samples	Limited to nectar volume > 0.1 μl. Each examination takes 1–3 min
Spot on paper (50)	Easy and handy. Suitable for large samples. Nectar collected under any conditions and analysed later. Gives amounts of sugar in a known volume	Accurate only in range of 18 to 60% sugar (*Figure 4*)
TLC	Portable and easy to use under field conditions	Semi-quantitative
Paper chromatography	Quantitative	Needs laboratory facilities
GC and HPLC	Very accurate also for small amounts of nectar	Need suitable equipment at the laboratory

ii. Sugar ratios

Baker and Baker (31), define a 'balanced' nectar as containing equal quantities by weight of sucrose, glucose, and fructose and then the sucrose: hexose ratio is 0.5. If there is as much sucrose as the combined weights of the hexoses, the ratio will be 1.0. Nectar with a sucrose:hexose ratio of less than 0.5 is regarded as 'sucrose poor' (or 'hexose rich'); nectar with a ratio of 0.5 or more (but less than 1.0) is 'sucrose-rich'; nectar with a sucrose:hexose ratio of 1.0 or greater is 'sucrose-dominated'. The main application of this approach is to correlate the sugar ratio type with the pollinator type (3, 52).

3.4.2 Amino acids

Protocol 9. Histidine scale for evaluation of amino acids in nectar (53, 54 and A. Erhardt, pers. comm.)

Materials

- histidine (MW 191.7)
- ninhydrin (0.2% in acetone) or prepared chromatography aerosol spray
- stripes of about 2 cm × 20 cm of Whatman No. 1 filter paper

A. *Calibration scale*

1. Dissolve 3.9 mg/ml of histidine in 20% sucrose solution. This stock is now referred to on a concentration scale of 10.

2. Prepare a series of 50% dilutions (scale 9 contains half as much histidine, as scale 10, etc., down to scale 1).

3. Keep all the concentration series (1–10) in small vials in the refrigerator. (They may be kept for several weeks.)

B. *Examination*

1. Mark 11 circles (each 0.5 cm diam.) with a pencil along the midrib of several filter paper stripes (for example, 2 × 20 cm; *Figure 4*), leave equal space of 1 cm between them and large margins on both sides (one stripe for 'calibration scale' and the other stripes for different nectar samples.

2. Mark clearly the area that was touched with your fingers while holding the paper. Do not touch the area of the circles (because sweat contains many amino acids).

3. Put 1– μl of each histidine concentration (1 to 10) in the middle of each appropriately numbered circle and distilled water in number 11. This stripe will serve as the 'calibration scale'.

4. Put 1–2 μl nectar samples in each circle of the other stripes

5. Write down the necessary identification details on the filter paper stripe.

6. Let the 'calibration scale' and the sample(s) dry.

7. Keep each cut stripe in a separate glassine envelope; you may store them at room temperature.

C. *Staining*

1. Add 1 to 3 μl of ninhydrin to the histidine references and to the nectar spots of the samples exactly in the centre of each circle, or spray them gently and uniformly with ninhydrin.

2. Dry them at room temperature for 24 h or in 85 °C for 10 min.

3. Cut the 'calibration scale' longitudinally at the middle of the spots.

4. Compare the colour (violet) intensity of the nectar samples with the calibration scale along the cut (*Figure 4*). The cut in the calibration scale allows a direct comparison with the colour intensity of the nectar samples.

5. Some amino acids produce colours other than violet with ninhydrin, most obviously proline and hydroxyproline (yellow) and asparagine (orange-brown). On the whole the histidine scale provides a good estimate of α-amino acid, concentrations in a very convenient, cheap, and easy way.

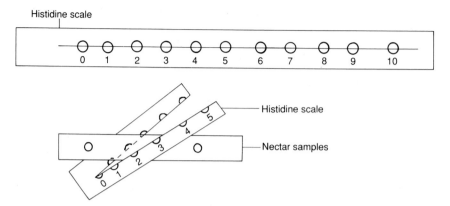

Figure 4. Comparison of nectar samples to the histidine scale to evaluate the amino acid concentration. (Courtesy of A. Erhardt.)

Protocol 10. TLC chromatography of amino acids in nectar (53, 54, and A. Erhardt, pers. comm.)

Materials

- dansyl chloride 1 mg/ml in acetone; UV lamp
- polyamide chromatography plates (15 × 15 cm) (e.g. Schleicher & Schuell or Chin-Chen Co.)
- amino acids standards; sodium bicarbonate; pH meter or pH test paper.

A. *In the field*

1. Spot the freshly collected nectar on the chromatography paper. After the nectar has dried, keep the paper in a tightly closed *dry* container until analysis preferably with some silica gel.

2. Use the same grade of chromatography paper for the samples as is used for the standards with which comparisons will be made.

3. If possible, make concentrated spots by repeated applications of nectar from the same source. Let them dry between applications.

4. Record the number and the diameter of the spots. This can be used to calculate the nectar volume (*Protocol 2*) and hence the total amount of amino acids present in a particular nectar sample.

B. *In the laboratory*

(i) *Identification of the amino acids*

1. Locate the exact position of the nectar spot on the Whatman paper with UV illumination (most nectar shows some fluorescence under UV light).

156

2. Cut out the spot and elute the nectar in 10–30 μl distilled water to cover the spot (depending on the size of the spot) adjusted to pH 8.6 with sodium bicarbonate. Disposable 6 mm × 50 mm culture tubes covered with paraffin are useful for this procedure.

3. Leave it in the refrigerator overnight.

4. Remove the resulting extract to a new tube and dry down in a vacuum dessicator.

5. Redissolve the amino acids (step 5 above) of the extract in distilled water made alkaline (pH 8.5) with 0.5 M $NaHCO_3$ (= 0.8 ml $NaHCO_3$ in 10 ml distilled water).

6. Adjust the amount of the pH water to produce an amino acid concentration of 100–500 pmol/μl. Then use equal quantity of dansyl chloride (1 mg/ml acetone).

7. The dansylation is complete when the colour fades (usually up to 1 h). Dry this extract until it is needed. (If the sample is also needed for analysis of the sugar composition, it should be not stored for more than one day, because dansyl chloride causes slow breakdown of the sugar.)

(ii) *Chromatography*

1. Redissolve the extract in 3 ml of chloroform:methanol:acetic acid (7:2:2 v/v/v) to get rid of the sugars which form a sticky mass at the bottom of the tube and which can afterwards be dissolved with distilled water and be used for determination of the sugar composition.

2. Spot the extract on to micropolyamide plates (squares of 5 cm), use finely drawn-out micropipettes keeping the spots as small as possible (< 2 mm).

3. When dry, chromatograph the plates in small beakers with the first solvent: formic acid:water (98.5:1.5 v/v).

4. At the end of this run, examine the plate under UV light (366 nm) to check the run. Dansyl-OH will have a bright-blue fluorescence about half way up the plate. If not, run the plate a second time in the first solvent.

5. Run the second solvent system, benzene:glacial acetic acid (9:1 v/v) in the direction of a right angle of the first. After drying rerun in system in the same direction.

(iii) *Tracing the acids*

1. Trace the actual distribution of the acids according to *Figure 5*.

2. Use acetate tracing paper (frosted on one side).

3. Detect the acids under UV illumination at 366 nm and then under 254 nm wavelength illumination.

As the amount of the fluorescence is proportional to the amount of the amino acids present, it is useful to produce a scale (using known amounts of

Protocol 10. *Continued*

dansylated amino acids) for estimating quantities present. For more precise quantification, a filter fluorometer (e.g. Turner, Model III) equipped with an automatic TLC scanning attachment can be used.

Notes

(a) Use only freshly harvested nectar from covered unvisited flowers. Insects may contaminate the nectar with their saliva and add amino acids (55). Avoid dropping any pollen into the nectar; it contains amino acids (56, 57).

(b) The method of Baker and Baker allows samples to be collected from remote places and analysed later (up to 2 years).

(c) Amino acids contents may change with flower age (58). Try to collect samples from different flower life-span stages separately for comparison to avoid possible misinterpretation.

(d) Comparison of the methods for evaluation of amino acids in nectar is presented in *Table 8*.

Table 8. Comparison of methods for evaluation of amino acids in nectar

Method	Advantages	Disadvantages
Histidine scale	Easy to handle under field conditions especialy in remote places. Instant results. Suitable for large sample	Semi-quantitative
Chromatography	Easy collection under field conditions in remote places and analysed later. Could be carried out in almost every lab	A lot of work in the lab
Amino acid analyser HPLC	Very sensitive and accurate quantitative method. Automatic analysis	Needs expensive equipment. Suitable only for fresh nectar

3.4.3 Lipids

Protocol 11. Lipid identification in nectar (Irene Baker, pers. comm.)

Materials

- Whatman No. 1 paper
- 1% osmic acid in water; Sudan III or Nile blue

Method

1. Spot the nectar on Whatman No. 1 paper.

2. Add a drop of 1% osmic acid in water.

3. Dry with cold air. A brown to black colour indicates the presence of lipids.

4. Read the colour shortly after adding the acid as other compounds react over time.

5. Parallel to this examination add a droplet of Sudan III or Nile blue to a fresh nectar sample on a glass slide. Oil globules in the nectar will be stained.

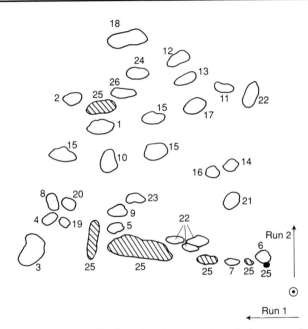

Figure 5. The amino acid distribution after the second run in benzene:glacial acetic acid (×2.3). (After Irene Baker, courtesy of A. Erhardt.)

1 Alanine	15 Methione (3 spots)
2 ß-Alanine	16 Ornithine
3 p-Amino butyric acid	17 Phenylalanine
4 Arginine	18 Proline
5 Asparagine	19 Serine
6 Cysteine	20 Threonine
7 Cystic acid	21 Tryptophan
8 Glutamine	22 Tyrosine (3 + 1 spots)
9 Glutamic acid	23 o-Tyrosine
10 Glycine	24 Valine
11 Histidine	25 Dansyl derivates (blue, 5 spots)
12 Isoleucine	26 p-Amino butyric acid
13 Leucine	⊙ Origin
14 Lysine	

3.4.4 Nectar pH

Any sensitive pH test paper is suitable for this examination. In the literature there is an almost total disregard of this aspect. In several flowers there are considerable changes in pH level which usually occur in parallel with accumulation of glucose, possibly as the results of microorganism activity. There is a need for investigation of changes in pH value relative to the other nectar variables and pollinator activity (but see ref. 58).

3.4.5 Other nectar constituents (Irene Baker, pers. comm.)

i. *Spot staining methods*
(a) Phenolics
  - Using the Folin–Ciocalteu's phenol reagent (Sigma F 9252). The reagent as purchased (2.0 N) is diluted to 0.5 N. Cover the nectar spot with the reagent (it is important to cover more than the spot, so that the background can be determined), and allow to dry. Add a drop of 20% Na_2CO_3. Phenolic compounds produce a blue colour which is deeper than the background. This stain is useful as a quantitative stain—provided the standard is clearly stated. Different phenols give different staining intensities!
  - Using *p*-nitroaniline. This uses three solutions:
    i 0.15 g p-nitroaniline in 4.5 ml conc. HCl + 95 ml water
    ii 5% $NaNO_2$
    iii 10% Na_2CO_3 (or Na acetate)
    To use: add about 10 drops (*i*) to 1 drop of (*ii*) (cold), add preferably about 6 drops of (*iii*) to bring up the pH (more of (*iii*) will cause the mixture to become cloudy very quickly. When this happens—discard.) Add it to a drop of nectar. Most phenols produce yellow and brown colours and occasionally red or blue.

(b) Alkaloids: Dragendorff's reagent (Sigma D 7158) is the one most commonly used. This detects heterocyclic N_2.

  Formula:
  Solution A: 0.85 g basic bismuth subnitrate in 50 ml 20% acetic acid
  Solution B: 20 g KI in 50 ml water

  Mix equal quantities of the two solutions and store in the refrigerator in a dark bottle.

  To use, dilute 1 drop of the stock with 10 drops of cold 20% acetic acid. Heterocyclic N_2 compounds give an orange to red colour on a yellow background. If such compounds are only in low concentration they may show up later when the background colour slowly fades.

(c) Cardiac (steroidal) glycosides—the Kedde reagent.

160

Solution A: 3,5-dinitrobenzoic acid is dissolved in MeOH to give a 2% solution

Solution B: 2 N KOH

Mix equal quantities just before using. A blue-purple colour is produced when these compounds are present.

(d) Organic acids (including ascorbic acid): This uses a 0.1% solution of the sodium salt of 2,6-dichlorophenol-indophenol in EtOH. Organic and keto acids become dark red on a blue background. If this colour lightens, these are reducing agents; for example, ascorbic acid will go white immediately. α-keto-glutaric acid turns a greyish colour. (On prolonged standing, the colour is also reduced by glucose and fructose.)

(e) Proteins: Cut out spot of nectar on Whatman No. 1 paper to be tested and place into 0.1% solution of bromophenol blue. Allow a minimum of half an hour for staining. Rinse in 3 changes of 5% acetic acid. The spot will be green at this pH. If it is now transferred into water made slightly alkaline it will become blue. This can be made half-quantitative with parallel calibration with albumin from bovine serum (Sigma A 7030).

References

1. Simpson, B. B. and Neff, J. N. (1983). In *Handbook of Experimental Pollination Biology* (ed. C. E. Jones and R. J. Little), p. 142. Van Nostrand Reinhold, New York.
2. Simpson, B. B. and Neff, J. N. (1981). *Ann. Missouri Bot. Gard.*, **68**, 301.
3. Baker, H. G. and Baker, I. (1983). In *Pollen: Biology and Implications for Plant Breeding* (ed. D. L. Mulcahy and E. Ottaviano), p. 43. Elsevier, New York.
4. Baker, H. G. and Baker, I. (1983). In *Handbook of Experimental Pollination Biology* (ed. C. E. Jones and R. J. Little), p. 117. Van Nostrand Reinhold, New York.
5. Williams, N. H. (1983). In *Handbook of Experimental Pollination Biology* (ed. C. E. Jones and R. J. Little), p. 50. Van Nostrand Reinhold, New York.
6. Vogel, S. (1963). *Akad. Wiss. Lit. (Mainz) Abh. Math-Naturwiss. Kl. Jahrgang*, 1962, 539.
7. Vogel, S. (1978). In *The Pollination of Flowers by Insects*. (ed. A. J. Richards), p. 89. Academic Press, London.
8. Vogel, S. (1983). In *Encyl. Plant Physiol.* (New Series), **12c**, 559.
9. Kevan, P. G. and Baker, H. G. (1983). *Annu. Rev. Entomol.*, **28**, 407.
10. Kevan, P. G. and Baker, H. G. (1984). In *Ecological Entomology* (ed. C. B. Huffaker and R. L. Rabb), p. 607. John Wiley, New York.
11. Baker, I. and Baker, H. G. (1979). *Am. J. Bot.*, **66**, 591.
12. Colin, L. J. and Jones, C. E. (1980). *Am. J. Bot.*, **67**, 210.
13. Loper, G. M. and Cohen, A. C. (1982). *Am. Bee J.*, **122**, 709.
14. Petanidou, T. and Vokou, D. (1990). *Am. J. Bot.*, **77**, 986.

15. Gentry Instruments, Inc. (1977). *Manual of New Improved Microbomb Calorimeter* (Pat. No. 3.451.267), Aiken.
16. Elias, T. E. (1983). In *The Biology of Nectaries* (ed. B. L. Bentley and T. E. Elias), p. 174. Columbia University Press, New York.
17. Percival, M. (1961). *New Phytol.*, **60**, 235.
18. Baker, H. G. and Baker, I. (1975). In *Coevolution of Animals and Plants* (ed. E. Gilbert and P. H. Raven), p. 100. University of Texas Press, Austin.
19. Baker, H. G. (1977). *Apidologie*, **8**, 349.
20. Baker, H. G. (1978). In *Tropical Trees as Living Systems* (ed. P. B. Tomlinson and M. H. Zimmerman), p. 57. Cambridge University Press.
21. Baker, H. G. and Baker, I. (1982). In *Biochemical Aspects of Evolutionary Biology* (ed. N. M. Nitecki), p. 131. University of Chicago Press, Chicago.
22. Lütge, U. (1977). *Apidologie*, **8**, 305.
23. Dafni, A. (1984). *Annu. Rev. Ecol. Syst.*, **15**, 253.
24. Gilbert, F. S., Haines, N. and Dickson, K. (1991). *Funct. Ecol.*, **5**, 29.
25. Zimmerman, M. (1988). In *Plant Reproductive Ecology: Patterns and Strategies* (ed. J. Lovett-Doust and L. Lovett-Doust), p. 157. Oxford University Press.
26. Baker, H. G. and Baker, I. (1973). In *Taxonomy and Ecology* (ed. V. H. Heywood), p. 243. Academic Press, London.
27. Cruden, R. W., Hermann, S. M., and Peterson, S. (1983). In *The Biology of Nectaries* (ed. B. L. Bentley and T. E. Elias), p. 80. Columbia University Press, New York.
28. Käpylä, M. (1978). *Ann. Bot. Fennici*, **15**, 85.
29. Willson, M. F., Bertin, I., and Price, P. W. (1979). *Am. Mid. Nat.*, **102**, 23.
30. Tepedino, V. J. and Parker, F. D. (1982). *Env. Entomol.*, **11**, 246.
31. Corbet, S. A. (1990). *Isr. J. Bot.*, **39**, 13.
32. Corbet, S. A., Unwin, D. M., and Prŷs-Jones, O. E. (1979). *Ecol. Entomol.*, **4**, 9.
33. Corbet, S. A. and Delfosse, E. S. (1984). *Aust. J. Ecol.*, **9**, 125.
34. Inouye, D. W., Favre, N. A., Lanum, J. A., Levine, D. M., Meyers, J. B., Roberts, M. S., Tsao, F. C., and Wang, Y. Y. (1980). *Ecology*, **61**, 992.
35. Baker, I. (1979). *Phytochem. Bull.*, **12**, 40.
36. Punchihewa, R. W. K. (1984). MSc. Thesis. Faculty of Graduate Studies, The University of Guelph, Guelph. (Mimeo).
37. Unwin, D. M. (1980). *Microclimate Measurement for Ecologists*. Academic Press, London.
38. Unwin, D. M. and Corbet, S. A. (1990). *Insects, Plants and Microclimate*. Richmond Publ. Co., Slough.
39. Willmer, P. G. and Corbet, S. A. (1981). *Oecologia (Berl.)* **51**, 67.
40. Cruden, R. W. and Hermann, S. M. (1983). In *The Biology of Nectaries* (ed. B. L. Bentley and T. E. Eilias), p. 223. Columbia University Press, New York.
41. Southwick, E. E., Loper, G. M., and Southwick, S. E. (1981). *Am. J. Bot.*, **68**, 994.
42. Bolten, A. B., Feinsinger, P., Baker, H. G., and Baker, I. (1979). *Oecologia (Berl.)*, **41**, 301.
43. Weast, R. C. and Astle, M. J. (ed.) (1978–79). CRC *Handbook of Chemistry and Physics*. CRC Press, Boca Raton, Florida.
44. Prŷs-Jones, and Corbet, S. A. (1987). *Bumblebees*. Cambridge University Press.

45. McKenna, M. A. and Thompson, J. D. (1988). *Ecology*, **69**, 1306.
46. Umbreit, W. W. and Burris, R. H. (1972). In *Manometric and Biochemical Techniques* (ed. W. W. Umbreit, R. H. Burris, and J. F. Stauffer), p. 161. Burgess Publishing Company, Minneapolis.
47. Gal, A. E. (1968). *J. Chromatog.*, **34**, 266.
48. Lewis, B. A. and Smith, F. (1969). In *Thin Layer Chromatography* (2nd edn) (ed. E. Stahl), p. 164. George Allen and Unwin, London.
49. Freeman, C. E., Bevcar, J. E., and Scogin, R. (1984). *Bot. Gaz.*, **145**, 132.
50. Pais, M. S. S. (1986). *Apidologie*, **17**, 125.
51. Baker, H. G. and Baker, I. (1979). *Phytochem. Bull.*, **12**, 43.
52. Baker, H. G. and Baker, I. (1990). *Isr. J. Bot.*, **39**, 157.
53. Baker, H. G. and Baker, I. (1973). *Nature (Lond.)*, **241**, 543.
54. Baker, I. and Baker, H. G. (1976). *Phytochem. Bull.*, **9**, 4.
55. Willmer, P. G. (1980). *Oecologia (Berl.)*, **47**, 270.
56. Gottsberger, G., Arnold, T., and Linskens, H. F. (1989). *Botanica Acta*, **102**, 141.
57. Gottsberger, G., Arnold, T., and Linskens, H. F. (1990). *Isr. J. Bot.*, **39**, 167.
58. Baker, H. G. (1985). *Aliso*, **11**, 213.

6

The plant–pollinator interface

1. Introduction

The title 'pollination ecology' covers studies of different levels of integration; for example, the individual pollen grain, the stigma, the individual flower or plant, population, species up to community level study. Pollinators, of course, may interact with each level to various extents.

Recognition of a pattern such as: intraspecific variation in flower characteristics (such as size, colour, number) or interspecific 'character displacement' (e.g. shifts in flowering time); or co-existence and co-phenology of plant species as well as of the pollinators, may supply the basic background for each pollination study. (But one may also have to consider the possibility of absence of community-level patterns; ref. 1.)

The possible processes which act on the individual plant, species, or among species (plants as well as animals) may include facilitation, competition, and deception, all of which are density-dependent at intra- as well as at interspecific levels. The mechanisms, in which pollinators are actively involved, leading to the observed patterns include: improper pollen transfer (e.g. restriction on the stigmatic space available or negative effect on conspecific pollen, and pollen loss) and/or competition (e.g. differential attraction and pollinator preferences which are caused by the neighbour's effect on visit frequency and/or on the forager pollen load; refs 2, 3, 4).

It is of prime importance to recognize that not all the plant species in a given pollination guild are equal at the same situation (5). In the same community (or species combination) the selective pressure operated by the pollinators on the plant species is, by no means, reciprocal to the influence exerted by the plants on the pollinator. Thus any interpretation of 'co-evolution' should be treated very cautiously (6, 7).

Even if the level of organization, patterns, processes, and mechanisms were well-studied, one has to bear in mind that other aspects of population ecology besides pollination, such as fruit maturation, seed germination, or vegetative propagation, are involved in the species fitness (5).

Pollination syndromes have to be regarded as a conceptual framework for studies, for example given typical 'hawk-moth flower' may be visited by bees which may contribute to more pollen transfer and seed production than the 'original' expected pollinator (8). The same species can be pollinated at different extents by different pollinators; for example, throughout the day, the season, or at various habitats. The critical point is the pollination efficiency of the given species under given circumstances in terms of seed production, the differential contribution to the next generation and its implications in microevolutionary terms.

While studying pollination ecology on a community level and as a part of the food web (9) we usually pay a little attention to the essential role which the animal senses play in the chain of events. For practical reasons, the sensory modalities are treated separately, although the animals appear to use inputs from different sensory modalities often simultaneously or in close succession (10).

2. Recording pollinator behaviour on flowers

Recording and analysing of pollinator behaviour on flowers is essential in any study concerning:

- the pollination syndrome in relation to forager activity
- distinction between pollinators and other visitors
- advertisement and visit frequency
- reward manipulation and utilization
- pollen dispersal, deposition, carry-over, and use
- pollination at the community level
- pollination efficiency
- pollination energetics
- pollinators in relation to weather variables
- resource utilization by the foragers

To quantify the various elements of animal foraging behaviour a sampling unit must be chosen which will allow standardization of the observations. A reproducible unit of observation permits comparison of the activity of foragers on various plant species, habitats, times, and seasons. Methods include observations along standard (or equal) transects or among a standard number of flowers (plants) during a given period of time (*Table 1*).

2.1 The transect method

In this method the observer walks at a constant pace along a known transect while recording the presence of the various visitors on the flowers/inflorescences in the sample. The data are recorded manually or on tape. In

Table 1. Methods of recording and evaluating forager activities on flowers

Method		Advantages	Disadvantages
Recording of the number of flower visitors present along a known transect (or sample of it) while walking at a constant pace		Easy, cheap, and handy method. Useful, especially in crops or open habitat studies with low plants. Allows the gathering of much data in a short time, especially while comparing co-flowering varieties or species. Useful in census of flower visitor presence/density/activity in relation to weather variables	Generally the collected data only apply to the presence of a known agent on the flower at a given time without any details on its behaviour. The results are subject to fluctuation of the flower density along the transect. The observer may distract or deter some of the visitors. Measurements under different weather conditions may result in different species as important visitor/pollinators. Not suitable under dense tropical conditions in which the observed species is highly dispersed
	Data tape recording		
Observation during a known period of time (about 10–30 min) on the same sample of flowers/inflorescences at different times during the day and the season		Easy, cheap and handy method. Enables extraction of the following data: Visit frequency and duration. Behavioural pattern (pollen collecting, nectar consumption, stigma touch, grooming behaviour, etc.), movement to other flowers or plants. Sporadic data from different dates and location may be pooled together	Subject to observer experience in recognizing the components of pollinator behaviour on flowers. Bird and bat pollinated flowers need long periods of observation (6 h and one night respectively). Less efficient when several foraging species are visiting simultaneously
		Suitable especially for high density of flowers when one observer cannot survey the whole plant or plot	
	Data tape recording		
Videotaping data		Enables analysis of the timing and duration of the following additional variables: flight directionality; flight speed, timing, and duration; enables a long observation on a single sample without losing details; documentation of simultaneous visits on the same flower group	Rather expensive equipment. Time-consuming analysis. Limited field of view in the focal plane. Battery limits observation duration in remote places
		The data analysis and results are not subject to the observer's subjective reaction. More accurate results, especially in time recording. Possibility of detecting more details in slow-motion viewing	

general, transects of 100 to 200 m (±3 m broad) are sufficient to achieve comparable results under fair weather conditions.

2.2 The fixed sample method

Observation is carried out during a standard period of time (10–30 min) on the same number of flowers or inflorescences. It is recommended that a portable tape-recorder and stopwatch should be used to gather all the required data. It is important to make a clear division between description of behaviour and its measurement. Later, a detailed transcription (which is time-consuming) is required. This allows a continuous observation without distraction for note taking and the recording of the duration of selected events. In a simpler version, when a tape-recorder is not available, the observer can prepare blank working tables in advance, the appropriate columns to be filled out in the field. In this method, the details regarding the frequency of each event (e.g. nectar sipping, pollen collection, stigma's touch, etc.) could be gathered later, in the laboratory. A portable microcomputer with data logger is very efficient for this purpose. If two or more foragers are present at the same time, a tape-recorder is more difficult to use than scoring behaviour on a prepared sheet.

2.3. Videotaping

Video-recording allows subsequent quantitative analysis of the data in the lab without losing details which may become later as having significant implications.

2.3.1 Video-recording—some practical hints

(a) Long inflorescences can be only partially observed in full because some of the inflorescences will not be included in the caught frame present at the focal plane; the same is true for deep tridimensional flowers. Videotaping is not ideal for these flowers or inflorescences. It is advisable to remove other flowers which are hidden in the canopy or are out of focus, and to remove other obstacles which interfere with direct observation, while knowing that this may alter the forager's behaviour.

(b) The background should contrast as much as possible with the subject so that the flower should stand out from the background. This can be achieved by video-recording against the sky or an open space or an artificial background.

(c) The video frame has limited space which is dependent on the distance from the object. For each object, the optimal distance from the object should be calibrated in order to record the necessary details dictated by the aim of the research.

(d) Television resolution is limited, and there is frequently a problem of identifying the pollinators (especially if they are less than ± 8 mm). This may be overcome by a parallel sampling of pollinators in correspondence with timed photography.

(e) When working with marked pollinators, it is almost impossible to read the numbers on the tags in the videotape. A coloured spot (preferable bicoloured) is more practical, although its detection depends on the angle and the speed of the pollinator in relation to the camera.

Note

Recording of pollinator visits on a flower should be accompanied with the following simultaneous (or nearly so) procedures:

(a) Measurement of the microclimate around the flowers and ambient weather conditions.

(b) Sampling of the pollinators for further identification and/or pollen analysis.

The field data gathered will later be converted into comparable tables for further analysis. Quantification of the frequency and the duration of each variable for each pollinator allows comparative analysis of the behavioural patterns of the different agents. Observations at different times, locations, and plant densities are subjected to different floral resources availability.

3. Analysis of foraging behaviour

A study of behavioural patterns of the foragers on flowers is the essential background to every study in pollination biology. Generally speaking, it is not a goal *per se*, but it is studied in relation to other aspects (e.g. breeding system, advertisement and attraction, reward structure, pollen placement and displacement, etc.).

The level of observation and analysis (*Tables 2* and *3*) of pollinators is chosen according to the specific research hypothesis and goal. Thus, it is impossible to present a detailed protocol; rather, general guide-lines with reference to other sections in this volume will be provided. It is self-evident that the means of any particular study should be chosen according to the forager type (birds, large insects, etc.) and its behaviour. The various methods of recording forager activities on flowers (*Table 1*) are the basis for analysis of the forager behaviour. For convenience, forager behaviour should be analysed in sequential stages: before landing on the flower, in the flower, in the individual plant, among conspecific plants, and among heterospecific plants. Each level of analysis may involve different facets of flower pollination and may have different implications with respect to other aspects

Table 2. Methods for evaluating forager activity and density in relation to flowers

Variable	Method	Advantages	Disadvantages
Index of visitation rate (11)	No. of total visits at the observation period/no. of available flowers at this period	The result is independent of the total no. of flowers per plant	Difficult to count large numbers of small flowers (e.g. Apiaceae). Many visitors do not respond linearly to patch size
Visitation rate (12)	No. of visits/flower × hours	Standardization of observation for different floral densities and various durations of observations	Non-visited flowers also have to be included
Effective visitation rate per flower per time unit (V) (13)	$V = (A \times N)/C$ when: A = no. of visitors per plant per time unit N = no. of visited flowers per visitor C = no. of flowers per plant	Standardization of observation for different plant size	Needs very detailed observations or videotaping for exact timing measurements
Visitation rate (VR) (14)	$$VR = \frac{FT}{HT \times FN}$$ FT = Foraging time per hour HT = Species-specific handling time of the flower FN = No. of flowers observed	Allow to compare various foragers which have different behavioural patterns	Needs detailed observation
Attractiveness index (12)	No. of visitors/available flowers/time unit	Consider the density of various plant species. Allow comparison between the attractiveness of co-flowering species	Visitors are not necessarily pollinators. A species may attract less foragers out of which the efficient pollinators are more than in other species
Foraging rate (= pollination efficiency) (15)	No. of flowers visited/time unit	Rapid method, suitable for monitoring pollinators in crops	Flower densities are neglected
Visitation rate (16)	Nf = No. of flowers Ng = No. of visitors Nv = No. of visitations Vf = visitations/flowers (Nv/Nf) Vf/Gf = Ratio of visitor and visitation rate $Gf = Ng/Nf$ = visitors/no. of flowers	Vf can be calculated for each flower type. Gf is a good measure for visitation magnitude in the single flower	If the visitors are not marked, the same individual may be counted more than once

Table 3. Components of the foraging behaviour elements and their applications

Level	Observed variables	Additional examinations	Applications in studies of:
Before landing on the flower (previsitation)	The ratio between approaches and actual visit per time/flower unit	Flower advertisement analysis: Colours Size and shape	Forager choice behaviour Deception in flowers Advertisement in flowers
	Flight pattern towards the flowers Direct flight Zigzag flight Flight route since last visit	Nectar guides, pollen guides Olfactory attraction	Role of visual vs. olfactory cues in attraction
Single flower	Reward harvesting: Pollen – Method of pollen collection Grooming behaviour Duration of gathering Movement pattern inside the flower	Flower life-cycle Pollen load on the stigma and on the forager Carry-over study Pollen amounts, and energetics Pollen tubes produced as consequences of the visit	Pollination energetics Pollination efficiency Pollination syndrome
	Nectar – Duration of visit per flower Movement pattern inside the flower	Nectar amount, constituents, and rhythms of secretion Microclimate Tongue length	Pollination energetics Forager nutrition and preference Pollination syndromes
	Stigma touch	Pollen load on the forager in relation to that of the stigma	Pollination efficiency Breeding systems
	Larceny: Flower piercing Secondary robbing Pollen thieving	Nectar amounts and constituents Pollen amounts	Pollination at the community level Consequences of flower crop size and plant dispersion

Table 3. *Continued*

Level	Observed variables	Additional examinations	Applications in studies of:
Inflorescence	Pattern of successive visits. No. of flowers visited. Duration per flower/ inflorescence. Type of reward collected. Percentage of stigmas touched	Intra-inflorescence floral phenology and reward structure	Breeding systems Foraging behaviour and pollination energetics Community studies
Plant	Intraplant movement directionality. Sequence of visiting. Number and duration of visits per flower/plant	Intra-plant phenology and the reward structure Nectar amount, concentration and distribution	Breeding systems Pollination energetics Foraging behaviour and energetics
Conspecific plants	Flight distance between two successive plants Change in direction: The angular directional difference from one flower to the next in relation to the direction of approach to the first	Pollen load on the stigma Pollen carry-over Pollen flow Nectar amount, concentration and distribution	Intraspecific competition Pollination energetics Breeding system and the population's genetic structure Foraging behaviour and energetics
Heterospecific plants	Number of visits to each species per foraging bout and/or time unit	Pollen load on the stigma and on the forager	Interrelationships at the community level (competition, facilitation, deception)

such as: the attraction to the flowers, the choice of the particular flower, resource utilization, pollination efficiency and pollination energetics, competition, breeding systems, pollen flow, and population size.

It is practically impossible (and logically improper) to differentiate between forager behaviour and its consequences. An attempt to review the main components of forager behaviour in relation to their most significant implications in pollination studies is given in *Table 3*.

When studying forager behaviour, several precautions and procedures should be kept in mind for further analysis and interpretation:

- Each forager caught should be kept separately in a pollen-clean vial with a numbered label (to correspond with the observation records). When studying pollen loads on insects, kill the insect in its own vial, and not in the general killing-bottle, to avoid contamination.

- Keep a reference collection of all the plant species involved and their pollen; voucher specimens of plants as well as of the pollinators should be deposited in larger collections or in museums.

Cautionary notes on the analysis and interpretation of pollinator behaviour on flowers:

(a) Most pollen-collecting insects may collect pollen from several plant species, and nectar from others. A knowledge of the range of the flower species visited (e.g. by analysis pollen on the insect's body: see Appendix A1, *Protocols 1* and *2*) is needed to understand the type of relationship between the flower species and the visiting agent. This background knowledge may have a tremendous influence on the measurement and the interpretation of pollen carry-over, distance of travel, and rate of pollen collection, the visitor constancy, nectar consumption, and pollination chances. These data are of prime importance in any analysis of community level interrelationships (mutualism,. competition, deception, etc.).

(b) Flower-visiting species within single orders, families, or even genera often differ in their foraging behaviour, pollination efficiency, abundance, or frequency of visits. Pooling data of individual species into broader categories (e.g. genera, families, order, etc.) may ease statistical analysis, but may mask the real differences among the species and their actual role (effectiveness) in pollination (17, 18).

(c) The categorization of pollinators according to body size may make sense in a broad-scale survey (e.g. throughout the seasons in relation to the floral morphology (ref. 18)) but must be scrutinized carefully in each specific case. This is especially so in a harsh environment and/or with a scarcity of pollinators, where there is no correlation between the pollinator (at least insects) size and the flower dimensions (19, 20).

(d) The observation of pollinators should be carried out during the whole life-span of the flower (including nights if the flower remains open) and flowering period of the populations, and, if possible, on various individual plants and habitats, seasons, and years. A local short study may include only a small fraction of the pollinators and only a narrow angle of the whole syndrome. Seasonal and yearly fluctuations may affect pollinators' species diversity, their behaviour and abundance, as well as choice of flowers, especially along an ecological gradient.

(e) The activity levels of many pollinators (especially insects) is temperature-dependent. The rate of nectar consumption is dependent on the temperature and water balance of the pollinator which is also influenced the relative humidity, wind velocity, and radiation. Thus, any analysis of pollinator behaviour should take the possible effects of all environmental variables into consideration (see, for example, refs 21 and 22).

(e) All flower visitors, not only the most frequent or the most efficient, should be scrutinized as potential pollinators. A large number of 'less important' pollinators may contribute more than a small fraction of the pollination in relation to that of the 'more important' ones (especially those which fit into the pollination syndrome in size and behaviour, the 'legitimate pollinators').

4. Pollination efficiency

Pollination is considered in its narrowest sense, as 'the process by which the pollen grains leave the anthers and reach the target stigma'. Most of the criteria for the evaluation of 'pollination efficiency' are related to different quantitative aspects of the pollination event. While considering the role of pollination in reproductive success, as determined by the production of viable seeds, one must be aware of the various developmental and genetical obstacles which exist between the deposition of the pollen on the right stigma at the right stage, and the final rate of seed production. Thus, various authors use the term 'pollination efficiency' (or similar terms) to assess various stages (some of which partially overlap) of the pollen's *via dolorosa* from anthesis to seed production. Semantic problems aside, each criterion may be used to elucidate different aspects in the boundaries of its limitations (*Table 4*).

5. Flower constancy

Since Aristotle's time, it has been well known that honey-bees tend to confine their visits to the same plant species on each collecting trip. The tendency of a pollinator to restrict its visits to flowers of a single species or morph, has been referred to as *flower constancy* (37, 38, 39). In addition, the term *fidelity* is

Table 4. Main definitions and criteria to evaluate 'pollination efficiency' (the plant's viewpoint)

Definition	Method	Advantages	Disadvantages
Pollination intensity (23). The number of pollen grains deposited on a stigma after a single visit	Exposing of virgin flower to one single visit and counting the pollen grains on the stigma	Convenient procedure to evaluate the relative pollen load contributed by different pollinators, also at various stages of the flowers' life-span	Pollen loads on the stigma do not necessarily relate to the chances of seed production
Relative efficiency (24). The amount of pollen transferred to the flower by a given pollinator type	Collecting the flower visitors, cleaning, and counting of all the pollen grains	Useful method for a comparative large-scale study at various places and times	There is no consideration of forager size, pollen viability, or its location on the body, in relation to its chances of reaching the stigmatic surface. In SI systems, no information on pollen compatibility is considered
Pollination efficiency (PE_i) (25) $$PE_i = \frac{(Pi - Z)}{(U - Z)}$$ Pi = Mean no. of seeds per flower of a plant population Z = Mean number of seeds per flower of a population receiving no visitation U = Mean no. of seeds set per flower of a population receiving unrestrained visitation	Exposure a virgin flower to one visit, re-cover and harvest the seed set for comparison seed set with production under nets and with free unlimited visits	Considers the final product of the reproductive success regardless of limitations or fertility problems at the various stages of the process Useful for comparison of various pollinators in various habitats, seasons, etc., regardless of their relative abundance	The U value is subject to seasonal fluctuations, ovule number, pollination limitation, and/or competition for pollinators. Artificial cross-pollination is needed as a complementary control to assess the potential seed production
Absolute pollination efficiency (26). The relation between pollen removal to pollen actual deposition	Measuring of the total pollen grain removal divided by the pollen amount delivered to stigmas of compatible recipients	Takes into consideration the actual percentage of pollen transferred to the stigma	Elaborate technique. Needs previous knowledge of the breeding system and the dye particles behaviour. No information on the fate of the pollen on the stigma

Table 4. *Continued*

Definition	Method	Advantages	Disadvantages
Pollination efficiency: Percentage of touched stigmas out of the total visited flowers (27)	Direct observation of visitor activity on the flower	This parameter makes it possible to differentiate between visitors and pollinators and to compare efficiency of various co-visiting agents	A stigma touch doesn't necessarily result in pollen deposition. Less meaning if there are several stigmas per flower. Pollen may not be deposited on every stigma touched
Pollinator efficiency: Percentage of fruit set at a given period of anthesis (27)	Exposing virgin flowers for a given time when different pollinators are active (e.g. day vs. night)	Useful in comparison of the relative efficiency of different pollinators active at different times during anthesis	Receptivity of the stigma may change during a long anthesis or at various times during the day. Simultaneous studies of stigma receptivity are required
Pollination effectiveness: Seed yield as a result of one visit of one pollinator (28)	Virgin flowers are exposed to one pollinator visit; the visited flowers are marked, recorded, and rebagged. Seed yield is recorded later	Allows comparison of efficiency of various agents in relation to seed production	Needs careful manipulation and is a tedious procedure. Needs large sample size
'Pollinator efficacy': The relative potential of a flower visitor species as a successful pollen vector for a given population of plants (29)	Ep = Pollinator efficacy: $Ep = S \times C \times P \times V \times D$ when: S = No. of stigmas contacted/no. of receptive flowers visited per plant C = Crossover probability (= 1/ no. of flowers visited per plant) P = Pollen carrying capacity (= average no. of grains on stigma-contacting body areas) V = Visitation rate (average no. of receptive flowers visited per min)	Enables comparison of individual plant pollination efficiency with that of the whole plant population	Some parameters are 'subjective scores'. It is almost impossible to extract all needed data to a high accuracy level. Needs laborious observations. Error in estimates is multiplied such that confidence limits on Ep may be very large

Criterion/Index	Measurement	Advantages	Disadvantages
'Pollination effectiveness' (30)	D = Relative field density (average no. individuals per receptive flower per min) $\dfrac{\text{No. of fruits}}{\text{Pollinated flowers}} \times 100$	Easy to obtain, practical criterion in crop science	
Pollination efficiency (31) (a) The proportion of pollen grains produced per flower which reach stigma of same species of plant (b) Number of conspecific grains which reach a stigma in relation to number of ovules to ber fertilized via that stigma (c) Reciprocal of pollen:ovule ratio	(a) Count average number of pollen grains per flower vs. average number of pollen grains deposited on stigma (b) Count the ratio of the conspecific grains on stigma and of fertilized ovules (c) $\dfrac{1}{P\!:\!O}$	(a) Evaluates actual pollen load in relation to the potential pollen grains produced per flower (b) Gives information on relative success of pollen grains on stigma (c) Easy to calculate, regardless of pollinators and their behaviour	(a) Gives rate of pollen arrival at target stigma regardless of whether or not it results in fertilization (b) Not all the pollen reaching the stigma is necessarily compatible Pollen and/or ovule competition may mask results. The criterion partially reflects the rate of success of germinating pollen grains (c) Reflects only one facet of breeding system and has nothing to do with pollinator presence, diversity, or behaviour
Pollination effectiveness (32). The correlation between the forager visitation frequency and average fruit and seed-set	Direct observation on visitation rate combined with monitoring fruit and seed set	Relates seed production to pollinators activity. Makes it possible to compare efficiency of different pollinators	Not all the visits terminate in pollen transfer and/or deposition
Index of pollen transfer effectiveness (33). Average no. of flowers visited per min. × ave. no. of pollen grains deposited on the stigma in a single visit	Monitoring of foragers visits in flowers and pollen load per visit on virgin stigmas	The index is species-specific and independent of the population size. Use to characterize different pollinators on the same plant species	No consideration of the pollen fate on the stigma Elaborate procedure

Table 4. *Continued*

Definition	Method	Advantages	Disadvantages
Pollination index (34)	P_{indx} = No. of seeds produced/ pollinated flowers	Easily reached criterion especially suitable for mass crop studies	Reliable only if there is a fixed number of seeds per fruit
Pollination efficiency (35)	$PE = \dfrac{\text{No. of developed fruits}}{\text{Total no. of flowers}} \times 100$	Needs only final enumeration of fruits; efficient index in crop pollination studies	Ignores the rate of flowers that were not pollinated at all, as well as pollinators behaviour density
Pollination intensity (*PI*): No. of conspecific pollen grains on the stigma. Pollination efficiency: average of *PI* from activity of a given pollinator population. Fertilization efficiency: average no. of seeds resulting from a given *PI* (36)	Counting no. of pollen grains on the stigma in relation to the seed yield, for a given pollinator	Use of all the criteria simultaneously allows a good evaluation of the pollen rate and the relative efficiency of the pollinators	Reliable only if the species is self-compatible. Needs a lot of field work

sometimes used to denote the same phenomenon (see ref. 40 for further discussion).

Floral constancy has been largely studied as a behavioural means of isolating reproductively plant species (37). Macior (41) was justified in pointing out that except for the behaviour of monolectic bee species, flower constancy must be viewed as a relative term, depending on the behaviour of the individual foragers.

Because there is a widespread use of the term 'constancy' to denote different (and sometimes even unrelated) phenomena, there is a need to clarify the terms. The chosen methods for each study on 'flower constancy' are derived from the viewpoint of the researcher and results which may lead to confusion and/or disagreement may originate from different interpretations of the same term.

Waser (42) suggests distinguishing constancy from two other kinds of specializations that might occur when there is access to several flower types. A pollinator may specialize because it has fixed floral affinities, as do oligotropic or oligolectic solitary bees (40, 43). This is referred to by Waser as *fixed preference* in comparison to *labile preference* in which a pollinator without fixed affinities may temporarily specialize on flowers that are abundant and rewarding.

Several behavioural patterns are included under the title of 'labile preferences' and treated differently by various authors. Heinrich (44), for example mentioned three types of 'floral constancy':

i. Different lines of honey-bees have different 'preferences' for the same plant species probably based on the relative tongue and corolla length.

ii. Foragers are forced to be relatively faithful because of the scarcity of other plant species and exclusion from an alternative food source. This situation is closely related to 'passive constancy' (45) which is a result of the spatial aggregation of flower species.

iii. Constancy on the basis of the individual, within a less constant species; which arises from recruitment and conditioning.

It seems to be practically impossible (especially in honey-bees) to distinguish clearly between these three situations.

The proximate mechanisms that promote flower constancy are (44):

(a) learning of shape, colour, and scent of the flowers;

(b) loyalty to a particular vicinity in foraging which restricts foraging area;

(c) learning of a certain time of day for foraging.

Flower constancy probably promotes efficiency in foraging. Even in a field of mixed and equally attractive and productive flowers, a bee can probably work more efficiently if its sense organs and behaviour are temporarily 'set' in a particular way (ref. 46, p. 140).

Waddington (47) noted that 'Fidelity of bees to flowers of a single species is simply one outcome of a whole array of possible behavioural patterns.' He also suggested that '. . . flower-constancy should be reserved for describing the singular situation of a bee visiting only a single plant species throughout a foraging bout; the term should not be used to describe generally the bee's choice pattern.' Instead of describing only one facet of the pollinator's possible movement repertoire, Waddington (47) suggests the term *floral-visitation-sequence* (F-V-S) as 'a general qualitative heading for describing a bee's flower visiting behaviour and as a heading to designate such studies . . . The important part of this term is 'sequence', for it is the actual sequence or order of visits to flowers that most completely describes a bee's behavior.' This view also includes heterospecific visits during the same foraging bout, a variation in pollinator behaviour which has been ignored because emphasis was placed on fidelity (48).

While flower constancy (in a broad sense, including *fixed* and *labile preferences*; *sensu* ref. 42) accounts only for a small fraction of the whole repertoire of possible behavioural patterns (47), research on other types of floral visitation sequences is almost a *terra incognita*. Several models were generated to predict pollinator behaviour (49, 50), but experimental evidence is still scarce.

5.1 Measuring floral constancy

One traditional method of measuring floral constancy (51, 52) levels is to analyse the pollen loads of the pollinators. The percentage of conspecific pollen is used as an index of flower constancy. Faegri and van der Pijl (ref. 40, p. 46) mentioned that 'pollinators that collect both pollen and nectar may exhibit separate degrees or even types of constancy for the two activities'. Following this view, Zahavi *et al.* (53) suggest that a pure pollen load demonstrates 'pollen constancy' rather than 'flower constancy'. This view is also shared by Free (ref. 39, p. 35), who noted that because most honey-bees and bumble-bees collect pollen on some trips and nectar on others, their overall constancy must be less than that indicated by examining pollen loads alone. This awkward aspect can be reconciled by also checking the pollen load in the honey-stomach load of the individual forager (ref. 39, p. 36). Direct observation of the individual forager reflects (*Table 5*) only a temporary situation and is more useful in any study of floral visitation sequence (FVS) in its broadest sense (47).

Bateman (58) proposed a constancy index that depends on the tendency for all transitions to occur between like flowers. A pollinator leaving any flower has access to the flower of types 1 and 2. When transition frequencies are known Bateman's index ranges from −1 (complete inconstancy; all transitions between unlike flowers) to 0 (random transitions) to +1 (complete constancy;

Table 5. Methods of assessing flower constancy and their applications

Method	Technique	Advantages	Disadvantages
Tracing individual pollinators	Direct field observation of marked individuals on successive visits to flowers on same foraging bout	Results reflect actual behaviour under field conditions and could be related to plant species density, reward structure, etc.	Observations record only a fraction of foraging bout. Results depend considerably on spatial distribution of plant species in the observation patch (54). Given highly non-random spatial mixing of flower types and complex pollinator movements, it is difficult to generate accurate expectations of encounter frequencies with different flowers (43)
Analysis of the pollen load on the forager	Analysis of the proportion of pure pollen loads (monospecific) to mixed pollen loads as an index of pollinator constancy	Easy to carry out and to quantify, especially for bees, and in relation to crops	Proportion of mixed pollen loads does not reflect abundance of pollen. Pollen from principal forage sources may comprise a minor component of a mixed load. Data are needed on movement among flowers of different species (ref. 39, p. 35)
		More or less accurate for pollen collection, especially honey-bee and oligolectic solitary bees.	Rarity of one pollen in a mixed load may reduce the implied constancy below its real level (ref. 39, p. 35). It measures only the constancy of pollen-gathering (53). There is no information on flower availability and movement between flowers (42)

Table 5. *Continued*

Method	Technique	Advantages	Disadvantages
Pollen analysis of pollinator nectar load	Harvesting the nectar load of an individual forager and analysis of pollen for purity	Allows the differentiation between pollen 'constancy' and nectar 'constancy' (ref. 39, p. 36)	No data on visit sequences (ref. 48). Pure load may be a result of a patch of only one flower type (ref. 39, pp. 34–5, and ref. 44). Limited to pollinators from which the nectar could be extracted. Not applicable to honey-bees because their honey stomachs are not empty when leaving the hive to forage (ref. 39, p. 36)
Artificial bouquet	Foragers are exposed to artificial arrangements of flowers under natural or semi-natural conditions. Videotaping of the foragers' movements	Control of species composition, relative distribution, and distances between flowers. Easy to record visitation sequences of individual pollinators	Exposure to an array experiment of pollinators under natural condition may reflect their previous experience (44), their behaviour not necessarily fitting the optimal foraging theory (55). The observed behaviour may exhibit only a small fraction of the pollinators' foraging bout. It is difficult to score all transitions or to achieve controlled accessibility with artificial bouquets (42). In several studies labile preference has not been distinguished from constancy (42)

Array experiments	Array experiments under field conditions in which flower species density and spacing are controlled. Videotaping of each individual forager is useful to prevent missing any movement	Arrays allow the simultaneous scoring of all possible transitions between flower types whose accessibility is controlled (42). The experiment's rigid design offers an equal chance for all kinds of transition between flowers carried out by the pollinator. The index is altered only by unequal changes in transition frequencies between like and unlike flowers in the same row or column and is insensitive to preference change (42). Easy to carry out and useful in studies of 'labile preferences' (*sensu* ref. 42)	The artificial experimental design is unlikely to occur under natural conditions. A high degree of dissimilarity between the two flower types may enhance constancy level (42) as well as differences in the flower complexity and available reward (refs 38, 41; ref. 56, p. 208; but see ref. 55). There is no previous knowledge of the forager's previous experience which may influence its choice
Experiments with models in enclosures	Bees are trained to one colour and later exposed to another colour. With and/or without reward	There is full control of 'flower' arrangement, frequency and reward. Learning curves can be obtained as a result of various manipulations (57)	Foragers may be influenced by previous experience (57)

all transitions between like flowers. This index is insensitive to preference changes; it is altered only by unequal changes in transition frequencies between like and unlike flowers in the same row or column (42).

Data to be used in forager constancy and preference determinations included a record of plant species for both the plants of departure and arrival for individual interplant flights. The fidelity index (FI) of Levin (59), FI = 1 − (observed frequency of interspecific flights/expected frequency inter-specific flights), can be used to estimate the degree of forager constancy for two plant species.

6. Pollination energetics

Nectar and pollen are the main calorific rewards in flowers. Nectar is a sugary liquid, and pollen is packed in small discrete particles which contain mainly starch or oil. Most studies on the quality of floral reward have been done on nectar because its calorific value and rates of consumption are easy to measure. Heinrich and Raven (60) pointed out the importance and the implications of pollination energetics in relation to the utilization of flowers as food resources.

Optimal foraging theory assumes that animals have evolved through natural selection to forage efficiently; if several behaviour patterns are possible, an animal should employ the behaviour that maximizes some currency, such as the rate of net energy intake during foraging, given other constraints acting on the animal into account (61).

The first question that is raised when attempting to discover whether animals do forage optimally is to define what an optimal solution might be. Clearly, we can only calculate a relatively crude approximation to the optimal solution. The reason for this is that animals maximize their inclusive fitness, and because we cannot measure fitness directly, it is necessary to make simple assumptions which might allow us to measure fitness according to some appropriate *currency* (ref. 62, p. 237). Experimental data (see, for example, refs 14 and 55) do not always corroborate the expected and theoretical models.

Another crucial issue is that nectar is not only a source of energy but also of water. Water balance is important as well as energy profit (63). One has also to consider that an optimality model based on particular assumptions and currencies is not designed to test the idea that animals are optimal (64) but only to test the assumptions inherent in that particular model (ref. 62, p. 238). Another critical point is that the rewarding value of a plant species is not only evaluated on the basis of nectar but also on the simultaneous presence of pollen (3).

The operating constraints on foraging pattern may be intrinsic or extrinsic. Intrinsic factors are the forager's behavioural reaction, sensory capabilities, physiological demands, and morphological characteristics. Extrinsic factors one may consider are the weather variable (especially temperature and humidity), nectar standing crop, and interaction with other foragers on the resources (e.g. competition).

Foraging costs are derived from studies on how much energy is needed for various activities. Energy expenditure is measured directly by the rate of oxygen consumption (a millilitre of oxygen consumption is equivalent to about 24 joules (ref. 65, p. 126). Honey-bees and bumble-bees in free flight consume about 80–85 ml O_2/g/h (45).

While considering the handling costs as an energetic expenditure it is evident that shorter durations of nectar probing per flower will save energy. For example, bumble-bees with longer tongues forage more quickly at flowers with long corollas than do short-tongued bees at the same flower (66). If the nectar is too concentrated, its higher viscosity demands more energy to pump it out, thus morphological research should also be accompanied with a parallel nectar study.

The usual method of measuring the rate of energy expenditure is to measure oxygen consumption, but this method is obviously difficult to employ on a free-flying bee. However, a good index for energy expenditure of a large bee can be obtained by measuring its temperature, weight, and passive cooling rate as it perches on a flower, and by timing the duration of its flights between flowers. The rate of oxygen consumption during flight, which had previously been obtained in the laboratory could be multiplied by the flight durations to provide an estimate of the energy expenditure of a flying bee (ref. 67, p. 97).

Experimental work on pollination energetics has progressed along two lines (47). Some studies constructed models and examined several assumptions using details about the behavioural and physiological mechanisms that mould foraging. The second line is to test in the field the predictions of the models on the energetic balance of foragers taken in the laboratory.

Any experimental study of pollination energetics should include the following components (*Table 6*; *Figure 1*).

(a) The reward amount, its calorific value, and spatio-temporal distribution.
(b) The animal foraging behaviour, movement patterns, speed, interplant flight distance, departure decisions, and angular deviations.
(c) The animal energy expenditure (flight and handling costs).
(d) The energetic balance of the forager in relation to the theoretical models.
(e) Pollen as a reward should also be considered.
(f) Nectar sugar constituents should be considered.

Studies on pollination energetics may deal with various aspects of forager behaviour with regard to the energetic gain at various levels from the animal

Table 6. Main components of foraging behaviour and pollination energetics studies

Components/parameters	Method	Advantages	Disadvantages
Foraging decisions: Floral choice and visiting sequence interfloral interplant interspecies	Direct observation under natural conditions or artificial arrays or bouquets or models—with respect to available reward	Exact monitoring of the forager route	Results depend on the situation field vs. laboratory, naïve vs. experienced animals. Presence of other foragers may influence the observed one
Approach to flowers: Rates of approaches to visits	Direct observation in relation to the available nectar or average standing crop	Less rewarding flowers may be avoided	Measure of the standing crop may overlook intrafloral variation in nectar quantity and lead to misinterpretation
Departure rules:	*Models:* Offering the forager the choice between models, vary in reward, height, depth, colour, etc., and monitoring of visits sequence, intervisit distances, and directionality	The nectar volume/conc. and the distances between flowers are controlled. A large sequence of visits may be recorded	Applicable only for animals that respond to training experiments. Results depend on degree to which animal is energy stressed, which may switch its mode of 'risk sensitive' behaviour from 'risk-prone' or *vice versa*
	Field observations: Measuring of the amount of nectar collected in a single flower by estimation of the time elapsed since the last visit in relation to the nectar production rate	Allows testing of the 'threshold departure' rule (68, 69) against 'probabilistic departure rule' (70) at intraplant and interplant level	Possible only in species with a constant and predictable rate of nectar production and if the forager emptied the flower; otherwise the nectar yield per visit cannot be evaluated accurately
	Study of nectar intake per flower in relation to the flight distance to the next visited flower or plant (71)	No previous assumption on equal reward in different flowers	
	Measuring the nectar along a gradient of vertical inflorescences vs. visits sequence and directionality (72, 73)	When there is a natural gradient of reward from the bottom to the top—a natural experiment of different nectar volumes on the same plant	The results from vertical inflorescences do not necessarily apply to other flower arrangements (71, 74)

Reward structure			
Spatio-temporal distribution of flowers	Phenological records and mapping of the individual flowers and plants	A basis for nectar availability considerations	Valid only for the time of recording
Timing and rate of nectar production	See Chapter 5, *Protocol 5*	The basic data for calculations of nectar consumption rate	Rate of nectar production may be changed as a result of nectar use
Nectar standing crop	See Chapter 5, *Protocol 2*	Easy to measure under field conditions	Valid only to the measurement situation
Nectar caloric value	See Chapter 5, Section 3.3	The basic data of every energetic calculation	Non-sugar components are neglected
Nectar concentration (density)	See Chapter 5, *Protocols 3, 4, and 5*	A crucial factor in ingestion rate	
Sugar constituents	See Chapter 5, Section 3.4	May be very important in the forager preferences	
Foraging costs:			
Flight costs Foraging range Flight distance Flight speed Relative time of flight and non-flight activities	Monitoring of marked individuals	Supply a general view on the individual searching area. Evaluate the percentage of flight time out of the foraging bout for energetic calculations	Flight costs are known only for very few foragers (see ref. 77 for review). Highly influenced by wind and temperature. The non-flight activities are not equal in energy expenditure
Handling costs Probing time Ingestion time	Measurements of tongue length handling time (videotaping) and floral tube length (66) in relation to nectar availability	Applied to any type of forager under field conditions. Studied especially in tubular flowers	Suitable especially for similar foragers (intra- and interspecific) which differ in their tongue-length. Affected by the forager's learning ability and experience
	Experiments with models of different depths, volumes and concentration (75, 76)	Inexperienced foragers can be used without previous conditioning	Suitable only for foragers that can be trained to models (social bees and birds)
Handling efficiency	Measurement of the time spent per flower in relation to the calorific gain. Obtained by comparison with non-visited flower under the same conditions. Volume intake per time unit in relation to joule gain in the same time	Allows quantification of the relation between probing time and nectar extracted during the same visit	Needs elaborate work. Depends also on nectar concentration and temperature which determines its viscosity

Table 6. *Continued*

Components/parameters	Method	Advantages	Disadvantages
Flight energetic expenditure	Laboratory experiments under controlled condition to measure metabolic rate as O_2 consumption per body weight per time unit (78, 79)	Flight energetic expenditure may be related to joules and sugar quantities needed	Flight expenditure may not be constant; depends also on temperature and wind speed, which may vary between experiments
			Using metabolic rates for energy budget, is limited to specific and often experimentally controlled activities; the metabolic rates may not correspond with those in natural conditions (45). An endothermic insect may vary its metabolic rate by an order of magnitude in a few minutes while perching (80)
Energy budget	Calculations of all the observed and expected energetic intake vs. expenditure to get the net gain	Energy balance allows the analysis of forager behaviour economics in relation to the optimality theories	Indirect costs of learning and experience, risk of predation, loss of time, and wear and tear are not considered (81). Problems of using the same currency. Any errors in measurement or estimation are multiplied greatly such that final figures may be very inaccurate

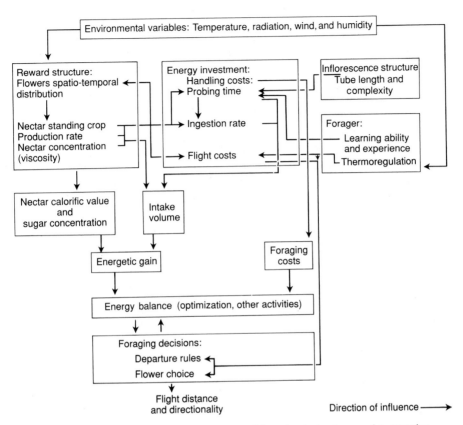

Figure 1. Interactions among the components of foraging behaviour and energetics.

viewpoint (individual foragers, animal species, and/or guild) as well as from the plant's viewpoint (reward structure and availability), and their interaction. *Table 6*, is an attempt to overview the main components of foraging behaviour and the energetic balance concerning pollination.

While studying pollination energetics it must be recognized that different constraints and 'goals' may result in very different 'decisions' taken by solitary and social foragers (e.g. bumble-bees, hawk-moths, humming-birds, many bees). Social but independent foragers that contribute to a collective store (e.g. bumble-bees); unsocial, interdependent foragers are 'cogs in a wheel' and contribute to a collective store (honey-bees, social meliponnines). 'Optimal foraging theory' may be inapplicable at least to the last group despite many attempts to enforce the theory (P. Feinsinger, pers. comm.).

Notes on the use of model flowers in pollination energetics studies (and in general):

(a) The reflectance spectrum curve of the models (including the stalks!) should be measured (see Chapter 4) in relation to the experimental background (82). This kind of measurement undoubtedly has great importance in evaluating possible signal-dependent learning curves (57).

(b) Certain foragers may prefer a specific form, size or colour; in energetics experiments there is especially a trend to use the preferred colour model to attract more foragers (see, for example, refs 47 and 48).

(c) In experiments which examine inherent reactions to floral cues (such as colour and shape) it is preferable to rear the foragers (83). Adult foragers caught in the wild have experience and are no longer naïve. Honey-bees also cannot be treated as naïve since there is an information exchange in the beehive. Different bees having identical preferences should show a large variety of age- or experience-related major and minor specializations under given assortments of different flowers (57).

(d) When several models (rewarding and non-rewarding as well) are involved in the same experiment, the position of the rewarding models should be changed in each run in order to reduce or eliminate site-specificity (55, 57).

(e) In using rewarding models, it may be difficult to train the foragers (especially honey-bees) with scentless ones; the models should be provided with a scent. Geraniol and peppermint extracts are usually used to enhance conditioning. If non-rewarding models are also involved, they should also be scented. When the forager learns to use the artificial flowers, the scent may then be removed (57, 84).

(f) The experimental model (system) should be washed carefully after each visit (if single visits are recorded) or after every foraging bout by the same forager. This is to avoid scent marks which may influence further visits (85, 86).

(g) While using flower models in studying foraging decisions the model should not offer an excess of reward; it may influence later decisions on search pattern as well as departure time (ref. 87, p. 91). Heinrich and Raven (60) have already mentioned that a single flower is not expected to satiate the energetic needs of the forager, and thus ensure visits to more flowers.

Furthermore, total caloric density should be realistic or should be varied to mimic natural situations in which foragers are or are not energy stressed, because very different behaviours may result (e.g. 'risk-prone' vs. 'risk sensitive' foraging, respectively).

7. Facilitation

Facilitation in pollination between two co-flowering or sequentially flowering species occurs when 'the presence of one species (or phenotype) increases visitation to another species at no cost to the first species' (88). This phenomenon has also been treated as 'mutualism' (89, 90) and 'cooperation' (91). Thus 'Müllerian mimicry' is also a kind of facilitation (88).

Because a high rate of visitation *per se* is not advantageous unless it results in increased reproductive success of the recipient flower (92, 93), this aspect must be borne in mind in any study which evaluates the adaptive nature of facilitation beyond the descriptive data on visitation rates and sharing of pollinators.

Heinrich (94) pointed out that flowers which offer only pollen are dependent on concurrent nectar sources for support of their potential pollinators (the reverse trend is also true!). This implies that a nectariferous species may share and support the pollinators of another species with a different flower structure and colour. If plants share pollinators, their interactions can be facilitative but their costs (via support) and benefits (via effectiveness gained) may be different, and the trade-offs between costs and benefits may be unequal (88).

Plant species which flower early in the season may support the initial pollinator populations which survive and reproduce and are then available to pollinate the later flowering species ('sequential mutualism', *sensu* ref. 90).

To demonstrate facilitation the following must be shown:

- both species share the same pollinator, habitat, and season
- reproductive success, at least of one species of plant, is higher when both species are co-blooming (or with sequential flowering) than it is when each species flowers without the presence of the other one

8. Competition

Waser (92) defines competition for pollination as 'an interaction in which co-occurring plant species (or phenotypes) suffer reduced reproductive success because they share pollinators'. A reduction in visitation rate in a competitive situation is not enough to demonstrate the effect of competition (95). There is a need to show the influence on components of reproductive success such as pollen transfer (83, 96); seed production (11, 12, 95, 97, 98, 99); and production of inviable or sterile hybrids (95, 100, 101, 102).

Competition through pollinator preference (103, 104) occurs when 'one plant species or phenotype within a species is somehow able to attract pollinators away from others and when this reduction in visitation lowers the reproductive success' (92). Others describe this type of competition as an 'exploitative competition' (12, 89).

Rathcke (88) suggests calling this situation simply 'competition', 'to avoid invoking a specific behaviour'. At any rate, both Waser (92) and Rathcke (88) refrain from adopting the zoologically derived term 'exploitative competition', to avoid misinterpretation.

Competition may also act by interspecific pollen transfer. It occurs when 'a pollinator forages without perfect preference in a mixture of plant species (or phenotypes) and causes pollen transfer between them, and when the resulting losses of pollen, receptive stigma surface, and effective pollination movements lower reproductive success' (92) but the evidence for it is scarce. Rathcke (88) defines this mechanism as 'Improper Pollen Transfer', (IPT), to include also genetically incompatible intraspecific pollen (although there are really very different phenomena functionally). Maybe one has to differ between interspecific IPT and intraspecific IPT, the second may occur even without the intervention of any external factor. Other authors relate to the zoologically derived term 'interference competition' (12, 89).

Improper pollen transfer may lower seed production through stigma clogging, and active disruption of the stigmatic surface, blocking the stylar transmitting tissues and eliciting flower abcsission or stigma closure (88, 93, 105). It also may lower the pollen carry-over and as a result influence the outcrossing rate as well as the distance of pollen flow (93) by production of inviable or sterile hybrids (88, 100, 101). In general, 'competition for pollination may act as a strong selective force on floral characters by reducing the quality as well as the quantity of pollen received by conspecific plants' (93). Theoretically, improper pollen transfer may lower seed production, resulting in the rapid exclusion of one of the two species (97).

Competition for pollination may promote evolutionary divergence among co-occurring species in resource use. Character displacement within and among species concerning flower morphology and colour, phenology, and reward structure has often been regarded as an outcome of competition for pollination *Figure 2*; refs 88, 92, 96, 98, 99, 106, but see refs 1 and 5.

Protocol 1. Competition for pollinators

Materials: See the different stages.

Method

1. Determine the phenological rhythms (see Chapter 1, *Protocols 2* and *3*) and the degree of overlap of the two (or more) putative plant species which are suspected to compete for pollinators.

2. Determine the spatio-temporal availability of the reward of all the plant species involved throughout the whole flowering season.

3. Monitor pollinator activity/frequency on each plant species (*Table 2*) and estimate the rate of pollinator sharing.

4. Determine the type of reward exploited by each pollinator on each plant species as well as its quality (see Chapter 3, Section 8).

5. Measure the pollination efficiency (see Section 4 above) of each pollinator species in the following situations:

- in the presence and in the absence of the competitor under natural conditions
- under controlled densitites of each plant competitor
- under nets without any pollination (see Chapter 2, *Protocol 1*)
- in comparison to hand-pollination.

6. When checking for Improper Pollen Transfer, also check the following additional aspects:

- the pollen load on the pollinator's body (Appendix A1, *Protocols 1*and 2) in relation to the spatial position of stigma(s) (of all the involved species).
- the pollen load on the stigmas of each species in relation to

  i pollen carry-over (Chapter 2, *Protocol 6*)

  ii pattern of flight movements and the sequence of visits between the competing plant species

- the influence of heterospecific artificial pollination on the seed set in comparison to a conspecific pollination

---

To demonstrate intraspecific competition for pollinators, one must show that reproductive success of an individual plant is reduced as a result of the presence of other co-flowering conspecific plants. Seed production as a result of hand pollination under nets should be used as a control to demonstrate the potential reproductive success when pollination is not limited.

While studying interspecific competition for pollinators, the situation (and its analysis) becomes more complicated. Measuring the relative attractiveness of each species (as revealed by the visitation index, *Table 2*) is not sufficient to demonstrate competition unless it is significantly related to pollination efficiency and seed production.

Any study should consider the relative density of each plant species throughout the season and between years, to eliminate any possible misinterpretations caused by the non-typical seasonal fluctuations. Quantification of the reward value (*Protocols 2* and *3* and Section 3.3.1), as well as of the advertisement magnitude (Chapter 4, Section 3.2), may clarify the mechanism by which one plant species attracts more pollinators than another. Plant species may compete for one or two pollinator species which they share

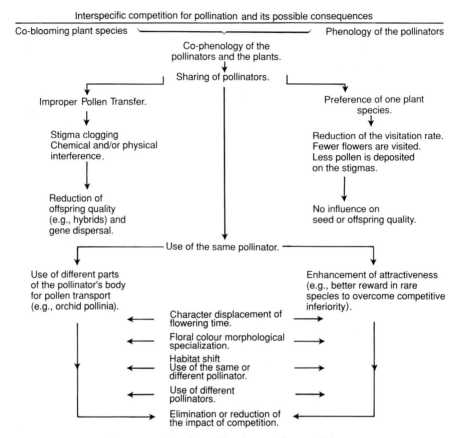

Figure 2. Interspecific competition for pollination and its possible consequences.

simultaneously or, in other parts of the season, they may compete for partially shared pollinators which may mask the effect of the competition.

The sharing of pollinators and the checking of the pollen load on the pollinator's body is not enough to demonstrate Improper Pollen Transfer (IPT) as the main factor in competition, unless the following precautions are considered:

- demonstration of the clogging of the stigma caused by heterospecific pollen as compared to a similar conspecific pollen load size
- the examination of the influence of a heterospecific pollen load on the seed production
- the presence of other plant species in relation to the reduction of the amount of conspecific pollen on the stigma

When the mechanisms of competition for pollinators are being considered, one should not exclude the potential possibility that both mechanisms, Improper Pollen Transfer and greater attraction, may act simultaneously in various combinations. It might be that species A is more attractive, but also has more pollen and will act by IPT on species B, or, for the sake of the argument, that species A is more attractive but suffers from the IPT of species B. Thus, a demonstration of *'exploitative' competition* does not automatically exclude IPT and vice versa. The situation may be even more complicated if more than two plant species are involved to different extents throughout the flowering season.

9. Mimicry and deception

Deception in pollination occurs when the flower attractants do not correspond with any type of floral reward; thus, the pollinator has no gain (food, shelter, etc.). Deceptive flowers are successful only when their forager is deceived twice, first to be loaded with pollen and secondly when the forager is attracted to another flower and deposit the pollen on the target stigma (107).

It is selectively advantageous for a forager to associate a suite of attractants (flower colour, shape, and scent) with reward. Some plant taxa 'exploit' the conditioning of foragers to certain floral cues and/or exploit naïve foragers, which are enticed by visual and olfactory cues prior to initial conditioning.

In terms of pollination ecology, Müllerian mimicry occurs where a number of similar-looking flowers of different species have evolved a common 'advertising style' which results in mutual benefits (ref. 108, p. 375). Müllerian partnership increases the 'effective density' of food resources to foragers and may increase the probability of pollination (109, 110, 111) without involving deception (112).

An 'integrated Müllerian and Batesian system' occurs when a minimum of three species are involved, two of which are rewarding (Müllerian type) and one (or more) non-rewarding (Batesian type), and all of which are similar in appearance (89, 113, 114).

Batesian mimicry occurs when a rare species providing no reward, mimics flowers of a more abundant species that does provide a reward (115). Two criteria are crucial in this situation: low frequency of the mimic (116, 117, 118) and a compensating reward by the more common model that subsidizes the system (119, 120).

Intraspecific Batesian mimicry includes imitation of male by female flowers (121) and automimicry when the model and the mimic are different individuals within the same species (122, 123).

The deceptive pollination systems are broadly divided into 'nutritive deception', in which the deception flower or part of it imitates a food source, and 'reproductive deception', in which the fraudulent flower simulates the substrate for oviposition or a female mate (see refs 122 and 124).

Table 7. Types of nutritive floral mimicry and deception in pollination

Mimic/ model	The same species	Two similar species	Several similar species	No model
Rewarding		Müllerian mimicry	Müllerian mimicry	
Deception	Automimicry. Imitation of male flowers by female flowers	Batesian mimicry	Integrated	Non-mimicry deception (125) = 'non-model mimicry' (107)
Deceptive + rewarding			Müllerian and Batesian system	

Generally, in nutritive deception the visual cues are of prime importance (*Table 7*; refs 119 and 122), because it involves conditioning and learning. As a result the system is subjected to avoidance of the mimics as a density-dependent process (116, 117).

Reproductive deception is involved in the mimicry by plants of the breeding sites of carrion and dung flies (sapromyophily), dung beetles (coprocantharophily), and fungus gnats (mycetophily), and in the imitation of female insects (pseudocopulation) (119, 122, 124, 125). In general, olfactory cues are decisive in reproductive deception, and they act on the innate behavioural patterns of the pollinators (119, 122). Thus, the pollinators are not subjected to learning and avoidance of the models (but see ref. 126), and the pollination efficiency is not density-dependent.

9.1 Batesian mimicry

To prove a case of Batesian mimicry between a rewarding species and non-rewarding one, the following needs to be demonstrated (127):

● The two species share a similar flower colour and pattern, habitat, and flowering season.

● There is a pollinator that visits both the non-rewarding species and the rewarding one.

● The non-rewarding species has better reproductive success as a result of its co-blooming with the rewarding species in comparison to appearing alone.

Protocol 2. Demonstration of Batesian mimicry

Materials: See the different stages

Methods

1. Quantify the phenological rhythms (see Chapter 1, *Protocols 2* and *3*) and the density of the model and its putative mimic.
2. Compare the floral morphology, the flower life-cycle, and the colour pattern (including the UV range (see Chapter 4, *Protocols 2* and *3*)) of both species.
3. Examine the type of the model's reward; demonstrate that the mimic is nectarless and that the pollen is not exploited as a food source, or that no other reward is present.
4. Monitor the pollinator's behaviour (*Tables 1* and *2*) to show a sharing of pollinators between the two plant species.
5. Check the pollen load on the common pollinator (Appendix A1, *Protocols 1* and *2*) and its exact location on its' body in relation to the spatial position of the appropriate stigmas.
6. Check the heterospecific pollen load on the stigmas of both species.
7. Compare the pollination efficiency and/or reproductive success (*Table 4*) of the putative mimic, with and without the presence of the model. Consider the various relative densities of the model and the mimic.

Because the success of nutritive floral deception is density-dependent, visits to mimic may be very rare at low density of the model, and the probability of observing and/or to catching a pollinator in action is low. An indirect method for identifying the models is the analysis of the pollen load on the stigma of the mimic. The deceived pollinator deposits conspecific pollen of the mimic as well as heterospecific pollen of the model(s) on the stigma, especially in orchids, which have a large and sticky stigma. Analysis of the pollen harvested from the stigmas of the mimic and a comparison with the pollen of other rewarding species may identify the model(s) and other species which indirectly subsidize the mimic. The next step is to catch large samples of foragers on the rewarding species, which are involved in the system, and which attract more foragers than the mimic. The collected foragers are then analysed for the presence of pollen of the mimic—as indirect evidence of the sharing of pollinators. In orchids, the number of pollinaria per forager allows estimation of the minimal number of visits in the deceptive species. A full comparison of the pollen spectrum on the mimic's stigma and on the shared foragers bodies will indicate which plant species are involved, even without observing any pollinator on the deceptive mimic plant. By this method it may be found that the model flower may be quite dissimilar to the mimic (at least to the human eye), and also how many models are involved during the whole flowering season of the mimic (128).

The relative density of the model(s) in relation to the mimic is crucial in terms of pollination success which dictates final seed production. At a high ratio of mimic to model, the forager may avoid the non-rewarding species as an outcome of learning. Thus, the relative densities of the mimic and the model, during the whole flowering season of the model in relation to seed production, are important in the analysis of the efficiency of deception.

The overlap between the model's and the mimic's flowering may vary between years and may involve several species of pollinators. Thus, at least several years of study are needed to cover the spectrum of all the pollinators and to examine the pollination efficiency of the system.

References

1. Feinsinger, P. (1987). *Rev. Chile. Hist. Nat.* **60**, 285.
2. Feinsinger, F., Tiebout, H. M., III, and Young, B. E. (1991). *Ecology*, **72**, 1946.
3. Jennersten, O. and Kwak, M. M. (1991). *Oecologia (Berl.)*, **86**, 88.
4. Campbell, D. R. (1985). *Evolution*, **39**, 418.
5. Feinsinger, P. (1987). *TREE*, **2**, 123.
6. Feinsinger, P. (1983). In *Coevolution* (ed. M. H. Nitecki), p. 282. University of Chicago Press, Chicago, Illinois.
7. Schemske, D. W. (1983). In *Coevolution* (ed. M. H. Nitecki), p. 67. University of Chicago Press, Chicago, Illinois.
8. Eisikowitch, D., Ivri, Y., and A. Dafni (1986). *Oecologia (Berl.)*, **71**, 47.
9. Shrivastava, U. (1985). In *Pollination Biology—An analysis* (ed. R. P. Kapil), p. 23. Inter-India Publications, New Delhi.
10. Schneider, D. (1987). In *Insects–Plants, Proc. 6th. Int. Symp. Insect–Plant Relationships* (Pau 1986) (ed. V. Labeyrie, G. Fabres, and D. Lachaise), p. 117. W. Junk, Dordrecht.
11. Zimmerman, M. (1980). *Ecology*, **61**, 497.
12. Pleasants, J. M. (1980). *Ecology*, **61**, 1446.
13. Andersson, S. (1988). *Oecologia (Berl.)*, **76**, 125.
14. Sowig, P. (1989). *Oecologia (Berl.)*, **78**, 550.
15. Richards, K. W. (1987). *Can. J. Zool.*, **65**, 2168.
16. Kevan, P. G. and Lack, A. (1985). *Biol. J. Linn. Soc.*, **25**, 319.
17. Herrera, C. M. (1987). *Oikos*, **50**, 79.
18. Shmida, A. and Dukas, R. (1990). *Isr. J. Bot.*, **39**, 133.
19. Kevan, P. G. (1972). *Can. J. Bot.* **50**, 2289.
20. Shmida, A. and Dafni, A. (1990). *Herbertia*, **45**, 111.
21. Corbet, S. A. (1990). *Isr. J. Bot*, **39**, 13.
22. Unwin, D. M. (1980). *Microclimate for Ecologists*. Academic Press, London.
23. Silander, J. A. and Primack, R. B. (1978). *Am. Mid. Nat.*, **100**, 213.
24. Ashman, T. L. and Stanton, M. (1991). *Ecology*, **72**, 993.

25. Spears, E. E. (1983). *Oecologia, (Berl.),* **57**, 196.
26. Galen, C. and Stanton, M. L. (1989). *Am. J. Bot.*, **76**, 419.
27. Dafni, A., Eisikowitch, D., and Ivri, Y. (1987). *Pl. Syst. Evol.*, **157**, 181.
28. Motten, F., Campbell, D. R., and Alexander, D. E. (1981). *Ecology*, **62**, 1278.
29. Sugden. E. A. (1986). *Am. J. Bot.*, **73**, 919.
30. Gudin, S. and Arena, L. (1991). *Sex. Plant Reprod.*, **4**, 110.
31. Richards, A. J. (1986). *Plant Breeding Systems*. George Allen & Unwin, London.
32. Montavlo, A. M. and Ackermann, J. D. (1986). *Am. J. Bot.*, **73**, 1665.
33. Herrera, C. M. (1990). *Oikos*, **58**, 277.
34. Visser, J. and Verhaegh, J. J. (1980). *Euphytica*, **29**, 379.
35. Mesquida, J. and Renard, M. (1978). *Proc. IVth Int. Symp. on Pollination. Md. Agric. Exp. Sta. Spec. Misc. Publ.*, **1**, 49.
36. Vassiére, B. E. (1991). Honey bees, *Apis mellifera* L. (Hymenoptera: Apidae), as pollinators of upland cotton, *Gossypium hirsutum* L. (Malvaceae), for hybrid seed production. Ph.D. thesis, Texas, A. & M., University, College Station.
37. Grant, V. (1950). *Bot. Rev.*, **16**, 379.
38. Laverty, T. M. (1980). *Canad. J. Zool.*, **58**, 1324.
39. Free, J. B. (1970). *Insect Pollination of Crops*. Academic Press, London.
40. Faegri, K. and van der Pijl, L. (1979). *The Principles of Pollination Ecology*, (3rd edn). Pergamon Press, New York.
41. Macior, L. W. (1974). *Ann. Missouri. Bot. Gard.*, **61**, 760.
42. Waser, N. M. (1986). *Am. Nat.*, **127**, 593.
43. Linsley, E. G. and MacSwain, J. W. (1957). *Wasmann J. Biol.*, **15**, 199.
44. Heinrich, B. B. (1975). *Annu. Rev. Ecol. Syst.*, **6**, 139.
45. Thompson, J. D. (1983). In *Handbook of Experimental Pollination Biology* (ed. C. E. Jones and R. J. Little), p. 451. Van Nostrand Reinhold, New York.
46. Michener, C. D. (1974). *The Social Behaviour of the Bees*. Belknap Press of Harvard University Press, Cambridge, Mass.
47. Waddington, K. D. (1983). In *Handbook of Experimental Pollination Biology*. (ed. C. E. Jones and R. J. Little), p. 461. Van Nostrand Reinhold, New York.
48. Waddington, K. D. (1983). In *Pollination Biology* (ed. L. Real), p. 213. Academic Press, Orlando, Florida.
49. Oster, G. and Heinrich, B. (1976). *Ecol. Monog.*, **46**, 179.
50. Waddington, K. D. and Holden, L. R. (1979). *Am. Nat.*, **144**, 179.
51. Betts, A. D. (1930). *Bee Wld.*, **2**, 10.
52. Betts, A. D. (1935). *Bee Wld.*, **16**, 111.
53. Zahavi, A., Eisikowitch, D., Kadman-Zahavi, A., and Cohen, A. (1974). *In Vᵉᵐᵉ Symposium International sur la Pollinisation*, p. 90. INRA Publ., Versailles.
54. Free, J. B. (1963). *J. Anim. Ecol.*, **32**, 119.
55. Wells, N. M. and Wells, P. H. (1983). *J. Anim. Ecol.*, **52**, 829.
56. Barth, F. G. (1985). *Insects and Flowers—The Biology of Partnership*. Princeton University Press, Princeton, NJ.
57. Heinrich, B., Mudge, P. R., and Deringis, P. G. (1977). *Behav. Ecol. Sociobiol.*, **2**, 247.
58. Bateman, A. (1951). *Heredity*, **5**, 271.
59. Levin, D. A. (1970). *Am. J. Bot.*, **57**, 1.

60. Heinrich, B. and Raven, P. H. (1972). *Science*, **176**, 597.
61. Pyke, G. H., Pullian, H. R., and Charnov, E. L. (1977). *Q. Rev. Biol.*, **52**, 137.
62. Davey, G. (1989). *Ecological Learning Theory*. Routledge, London.
63. Bertsch, A. (1984). *Oecologia (Berl.)*, **62**, 325.
64. Maynard-Smith, J. (1978). *Ann. Rev. Ecol. Syst.*, **9**, 31.
65. Kleiber, M. (1975). *The Fire of Life: An Introduction to Animal Energetics* (2nd edn). R. E. Krieger Publ. Comp., Huntington, New York.
66. Inouye, D. W. (1980). *Oecologia (Berl.)*, **45**, 197.
67. Heinrich, B. (1979). *Bumblebees Economics*. Harvard University Press, Cambridge, Mass.
68. Hodges, C. M. (1981). *Anim. Behav.*, **29**, 1166.
69. Hodges, C. M. (1985). *Ecology*, **66**, 179.
70. Cresswell, J. E. (1990) *Oecologia (Berl.)*, **82**, 450.
71. Kadmon, R. and Shmida, A. (1992). *Evol. Ecol.*, **6**, 142.
72. Pyke, G. H. (1978). *Theor. Popul. Biol.*, **13**, 72.
73. Pleasants, J. M. (1989). *Am. Nat.*, **134**, 51.
74. Klinkhamer, G. L. and Jong, de J. (1990). *Oikos,*, **57**, 399.
75. Waddington, K. D. and Gottlieb, N. (1990). *J. Ins. Behav.*, **3**, 429.
76. Harder, L. D. (1983). *Oecologia (Berl.)*, **57**, 274.
77. Heinrich, B. (1975). In *Coevolution of Animals and Plants* (ed. L. E. Gilbert and P. H. Raven), p. 141. University of Texas Press, Austin, Texas.
78. Heinrich, B. (1972). *J. Comp. Physiol.*, **77**, 40.
79. Bertsch, A. (1983). *Oecologia (Berl.)*, **59**, 40.
80. Heinrich, B. (1972). *J. Comp. Physiol.*, **77**, 65.
81. Waddington, K. D. (1985). *J. Insect Physiol.*, **31**, 11.
82. Kevan, P. G. (1983). In *Handbook of Experimental Pollination Biology* (ed. C. E. Jones and R. J. Little), p. 3. Van Nostrand Reinhold, New York.
84. Kugler, H. (1956). *Ber. Dtsch. Got. Ges.*, **69**, 387.
84. Frisch, K. von (1967). *The Dance Language and Orientation of Bees*. Belknap Press of Harvard University Press, Cambridge, Mass.
85. Beker, R., Dafni, A., Eisikowitch, D., and Ravid, U. (1989). *Oecologia (Berl.)*, **79**, 446.
86. Pham-Delegue, M. H., Masson, C., Etievent, P., and Azar, M. (1986). *J. Chem. Ecol.*, **12**, 781.
87. Gould, J. L. and Gould, C. G. (1988). *The Honey Bee*. Scientific American Library, a division of HPHLP, New York.
88. Rathcke, B. (1983). In *Pollination Biology* (ed. L. Real), p. 305. Academic Press, Harcourt Brace Janovich, Orlando, Florida.
89. Brown, J. H. and Kodric-Brown, A. (1979). *Ecology*, **60**, 1022.
90. Waser, N. M. and Real, L. (1989). *Nature* (Lond.), **281**, 670.
91. Thompson, J. D. (1981). *J. Anim. Ecol.*, **50**, 49.
92. Waser, N. M. (1983). In *Handbook of Experimental Pollination Biology* (ed. C. E. Jones and R. J. Little), p. 277. Van Nostrand Reinhold, New York.
93. Campbell, D. R. (1985). *Evolution*, **49**, 418.
94. Heinrich, B. (1977). In *The Role of Arthropods in Forest Ecosystems* (ed. W. J. Mattson), p. 41. Springer, New York.
95. Waser, N. M. (1978). *Ecology*, **59**, 934.
96. Campbell, D. R. and Motten, A. F. (1981). *Bull. Ecol. Soc. Am.*, **63**, 99.

97. Waser, N. M. (1978). *Oecologia (Berl.)*, **36**, 323.
98. Motten, A. F. (1982). *Am. J. Bot.*, **69**, 1296.
99. Kohn, J. R. and Waser, N. M. (1985). *Am. J. Bot.*, **72**, 1194.
100. Levin, D. A. (1972). *Am. Nat.*, **106**, 453.
101. Mulcahy, D. L. and Mulcahy, G. B. (1975). *Theoret. Appl. Genet.*, **46**, 277.
102. Thompson, J. D., Andrews, B. J., and Plowright, R. C. (1981). *New Phytol.*, **90**, 777.
103. Lack, A. (1976). New Phytol., **77**, 787.
104. Reader, R. J. (1975). *Can. J. Bot.*, **53**, 1300.
105. Waser, N. M. and Fugate, M. L. (1986). *Oecologia (Berl.)*, **70**, 573.
106. Pleasants, J. M. (1983). In *Handbook of Experimental Pollination Biology* (ed. C. E. Jones and R. J. Little), p. 375. Van Nostrand Reinhold, New York.
107. Dafni, A. (1986). In *Plant Surface and Insects* (ed. T. R. E. Southwood and B. E. Juniper), p. 86. Edward Arnold, London.
108. Proctor, M. and Yeo, P. (1973). *The Pollination of Flowers*, p. 375. Collins, London.
109. Heinrich, B. (1975). *Evolution*, **29**, 325.
110. Bierzychudek, P. (1981). *Biotropica*, **13**, 54.
111. Schemske, D. W. (1981). *Ecology*, **62**, 946.
112. Little, R. J. (1983). In *Handbook of Experimental Pollination Biology* (ed. C. E. Jones and R. J. Little), p. 294. Van Nostrand Reinhold, New York.
113. Ackerman, J. D. (1989). *Madroño*, **28**, 101.
114. Thien, L. B. and Marcks, B. G. (1972). *Can. J. Bot.*, **33**, 19.
115. Schemske, D. W. (1980). *Ecology*, **62**, 946.
116. Carlquist, S. (1979). *Aliso*, **9**, 411.
117. Macior, L. W. (1970). *Am. J. Bot.* **57**, 716.
118. Yeo, P. F. (1972). In *Insect–Plant Relationships*. 6th Symp. Entomol. Soc. London (ed. H. F. van Emden), p. 51. Blackwell, Oxford.
119. Vogel, S. (1975). *Ver. Detsch. Zool. Ges.*, **1975**, 105.
120. Nilsson, L. A. (1983). *Nord. J. Bot.* **3**, 157.
121. Wilsson, M. F. and Ågren, J. (1989). *Oikos*, **55**, 23.
122. Dafni, A. (1984). *Ann. Rev. Ecol. Syst.* **15**, 259.
123. Gilbert, F. S., Haines, N. and Dickson, K. (1991). *Funct. Ecol.* **5**, 29.
124. Ackerman, J. D. (1986). *Lindelyana*, **1**, 108.
125. Dafni, A. and Bernhardt, P. (1990). *Evol. Biol.* **24**, 193.
126. Paulus, H. F. and Gack, C. (1990). *Isr. J. Bot.*, **39**, 43.
127. Wiens, D. (1978). *Evol. Biol.*, **11**, 365.
128. Dafni, A. and Calder, D. M. (1987). *Pl. Syst. Evol.* **158**, 11.

7

Abiotic pollination

1. Wind pollination

Wind pollination (anemophily) is the dominant type of abiotic pollination, comprising some 95 to 98% of all known examples and prevailing in several families: Poaceae, Cyperaceae, Juncaceae, and within the orders of Amentiferae and Utricales. In many families of entomophilous plants there are a few members that have become anemophilous; for example, species of *Thalictrum* (Ranunculaceae), *Ambrosia* (Asteraceae), and *Fraxinus* (Oleaceae) (1).

The frequency of anemophily increases with latitude and elevation, being relatively low in tropical environments and very high boreal forests (2, 3, 4). There is an increase in the frequency of anemophilous species with elevation, both in tropical and temperate latitudes (2, 3, 4), and also in remote islands' floras and early successional stages.

Wind-pollinated species generally share common characteristics, not all of which are expected to be found in the same species. The typical features of wind-pollinated species may be summarized (1, 4–7) as follows: large quantities of dry pollen with smooth surface are produced which generally spread individually or in small groups (Reynolds number around 0.1 at a terminal velocity of 5 cm sec^{-1}; ref. 6); the stamens are usually large, borne on long filaments, and are well-exposed, often hanging or organized in catkins; pollen release is arrested till suitable conditions (i.e. warm and dry weather) prevail and are then dispersed rapidly and efficiently; there are few ovules per flower and large well-exposed and often finely divided and feathery stigmas; the perianth is generally much reduced or absent altogether; colour, scent, and nectar are frequently absent.

Sexes in separate flowers or separate plants frequently show a strong dichogamy and relatively close spacing of compatible plants; the structure and location of flowers and inflorescences on the plant maximize the pollen's entrainment in moving air—exposure of flowers above or outside the leaf mass, reduction of leaf surface, or flowering before leafing.

Factors controlling and maximizing wind-pollination efficiency are dependent, in addition to the plant's features, also on environmental factors and their interaction with the plant and the vegetation features (6, 8, 9, 10, 44; *Figure 1*).

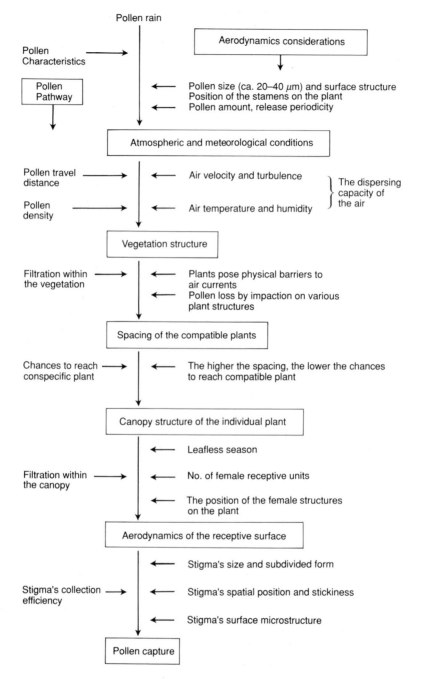

Figure 1. Factors controlling the airborne pollen pathway from its release until it reaches the receptive surface.

The vegetational structure in which anemophilous species dominate, is generally open and minimizes filtration of pollen by a non-stigmatic surface (2). Wind velocity needs to be in an acceptable range to ensure transport and minimize downwind dispersion (4, 8). Relatively low humidity and the low rainfall probability will enhance the chances of pollen arriving at its target (6, 8).

The aerodynamic environment around the ovulated organ is influenced by the speed and the direction of ambient airflow and by the geometry of reproductive and vegetative surfaces that obstruct air-flow (ref. 10, and earlier). This close environment influences the pollen trajectory towards the stigmatic surface (or the naked ovule). Wind-tunnel experiments accompanied with stroboscopic photography methods were used to visualize the behaviour of airborne pollen around the receptive surface in relation to its morphology (11, 12).

Several species are pollinated by wind as well as by insects (13, 14, 15) and exemplify intermediate characteristics especially with regard to pollen-kit stickiness (16, 17). In other cases predominantly animal-pollinated species also have a subsidiary wind-pollination, especially under unfavourable conditions or scarcity of pollinators (18).

Insects may collect pollen frequently from typical wind-pollinated species as solely a food source but without pollination, especially in dioecious species. Any demonstration of their possible role in pollination should follow Cox's postulates (19, p. viii). Even if insects are involved in the pollination of predominantly wind-pollinated dioecious species, the most efficient pollen-loaded vectors on the male plants are not necessarily the most efficient pollinators while visiting the female plants (Dafni, unpublished).

Protocol 1. Demonstration of wind-pollination

Materials

- fine mesh (1 × 1 mm) pollination bags
- air-proof pollination bags
- tags

Method

1. Cover samples (10 branches or plants, depend on the species dimensions) with fine-mesh bags, before flowering.
2. Cover a parallel sample with air-proof bags (each branch or plant bearing female organs in similar conditions to another branch or plant of the former treatment), before flowering.
3. Tag each branch or plant and note each one's position in the canopy, height above the ground, and/or the whole plant microhabitat (e.g. exposed vs. wind-protected plants).

Protocol 1. *Continued*

4. Keep the bags till the end of the fruiting season and count the fruit production rate in relation to the number of flowers covered in each treatment.

5. For hermaphrodite species check also the possibility of agamospermy and geitonogamy (see Chapter 2, *Table 7*).

6. If animals are also involved in pollination, cover another parallel sample of flowers with coarse mesh bags (the exact opening size depends on the observed pollinators).

The mesh of 0.25 × 0.25 mm (20) up to 1 × 1 mm (21) is regarded (depending on the species) as large enough to enable airborne pollen to penetrate the net and to reach the stigma but still dense enough to prevent pollinators. The fruit production rate under this net has to be regarded as a minimal value since the nets disrupt the air currents around the flowers and may thus change the stigma's collection efficiency (8, 10). If pollinators are definitely not involved there is no need to use any extra nets. The air-proof bags are used as a control and to evaluate (in hermaphroditic flowers) the existence of spontaneous self-pollination.

If, in addition to wind-pollination, animals are involved in the pollination process, the substraction of the results of fruit production under coarse-mesh from the total seed production under free pollination will give an estimate for the animal pollination rate.

1.1 Pollen traps

Pollen traps are extensively used in palynological studies (ref. 22; ref. 1, pp. 16–21). Many pollen samplers have been developed and are available commercially. Ogden *et al.* (22) survey not less than a dozen pollen traps with full descriptions and instructions. Each of these samplers or traps has its specific advantages and disadvantages (*Table 1*), most of which were designed to sample (as quantitatively as possible) the airborne pollen spectra rather than the pollen behaviour of isolated species.

For most of pollination studies of anemophilous species two main factors seems to be relevant:

- the specific pollen travel distance in relation to environmental conditions, plant and vegetation structure
- the pollen deposition rate on the stigmas.

While pollen travel distance can be clearly demonstrated, its chances to reach the stigma can only be estimated roughly and indirectly without quantification of the pollen density per air volumetric unit flowing over the stigma per time unit.

Many adhesives are used in pollen traps (ref. 22, pp. 74–75) especially for long-run sampling which demands resistance to adverse weather conditions. These adhesives are mainly based on Vaseline, silicon grease, and silicon oil. For short-term sampling Käpyalä's (24) mixture or polyvinyl–lactophenol was found satisfactory with later mounting in glycerol-jelly (see Appendix A2).

Protocol 2. Slide traps for airborne pollen

Materials

● Microscopic slides, glycerol-jelly (or Käpyalä's mixture or polyvinyl–lactophenol or gelatine–fuchsin)

Methods

1. Gently smear a fine layer (as uniformly as possible) of glycerol-jelly, Käpyalä's mixture, polyvinyl–lactophenol, or gelatine–fuchsin (see Appendix A2), on a sample of microscopic slides.

2. Leave the slides in a vertical position for several hours at room temperature to drain the excess liquid (if only glycerol is used).

3. Number the coated slides and store them in a tightly closed pollen-free (clean carefully in advance) container. Be careful to keep the slides separate.

4. Place the coated slides for 24 h (or any other given periods) at various distances and heights from the pollen source, at least six replicas for each set of conditions. Be careful to fix the slides horizontally to prevent pollen movement.

5. Place samples of coated slides near female (or hermaphrodite) flowers at various locations on the plant (six replicas for each similar condition); fix them vertically (as far as possible).

6. At the end of the sampling period collect each slide individually; do not touch the sticky surface. Separate the slide by tightening a rubber band around the margins of each one or with pieces of matchstick (of the same size as the slide width).

7. Refrigerate the sample till the examination.

8. Count the pollen number per area unit (at least 10 per each slide, the area size depending on the pollen density). If the pollen needs to be stained, gently add a small amount of methylene blue (or use pre-stained glycerol-jelly; Appendix A2). Be careful not to float the sticky surface; the pollen grains will move and may change the counting considerably.

Protocol 2. *Continued*

9. If the whole sample is used, wash the slides with absolute ethanol, put it into a container of known volume and count the pollen (Chapter 2, *Protocol 3*) or prepare it for acetolysis (Chapter 3, *Protocol 12*). Acetolysis is especially needed if too many other particles are stuck to the trap so that the pollen count is obstructed.

10. Identify the pollen by comparison to a reference collection and/or samples of the studied species.

The slides can also be smeared with Vaseline or agar (25). The advantage of using gelatine–fuchsin is the instant staining, but if the gelatine is too solid the pollen harvest may be considerably reduced. Gelatine–fuchsin is also extremely sensitive to changes in humidity and temperature causing variations in the retention efficiency (ref. 23, pp. 74–75). For short-term period (say, 24 h), however, it is quite satisfactory. For other adhesives and their properties consult ref. 22, pp. 74–75.

The pollen amount caught in various types of traps also depends (besides on wind velocity and turbulence) on the trap size, shape, and height from the ground level and its inclination (see ref. 1, pp. 15–19, ref. 22, pp. 49–70). A narrow surface is more efficient if trapping pollen from moving air than a broad surface of the same area (ref. 5 p. 263).

Collection efficiency is inversely proportional to the diameter of the collecting object; small objects collect more efficiently than large ones. This is due to the fact that large-diameter objects have proportionally thicker boundary layers of air surrounding them (8).

For pollination studies the main interest is the pollen travel distance and the pollen deposition rate on the stigma (number of grains per area unit per time unit). The common pollen traps (ref. 22, pp. 49–70) were moulded to measure the following parameters:

(a) the pollen content of the air, i.e. the number of grains in a static volume of air;

(b) the pollen deposition from the air, resulting from the number of grains passing through a cross-section per time unit;

(c) the pollen deposition from the air, resulting from the number of grains settling on a horizontal surface in a given time.

As far as sampling goes, this means pollen which is (a) immobile, (b) is horizontal, and (c) is in an integrated downwards movement. The three quantities are obviously related, but there is no constant, linear regression between them. Wind velocity and turbulence, precipitation, and surface character constantly change the interrelationships (ref. 23, p. 16).

Stigmas are irregularly shaped, bearing on movable parts of the plant and exposed to a constantly changing regime of wind velocity and turbulence.

Thus, pollen traps positioned at various parts of the plant and levels from the ground level would yield only relative data suitable for the particular conditions of the sampling. Extrapolations from the capture of pollen grains on the traps should be done with great care also, because of the specific size, the tridimensional shape of the stigma in relation to the pollen trap (15). As far as the identification of the airborne pollen grains and evaluation of the relative capture (at the sampling time) is concerned, the results from slide traps are quite satisfactory. Other traps which are used in aerobiological research (e.g. the Burkard trap and rotorod or rotoslide (ref. 22; *Table 1*) can be replaced by simple microscopic slides for most of the purposes in pollination study. The easy use and the possibility for many simultaneous replicas may compensate for more accurate mechanical traps.

The use of other pollen samplers in pollination biology is quite limited (but see ref. 11). General guide-lines about the most applicable methods are given in *Table 1* based on Ogden *et al.* (22), to whom the reader is referred for technical and methodological details.

Protocol 3. Evaluation of pollen 'scattering image' (ref. 13; ref. 26, pp. 2–5).

Materials

Stelleman–Knoll apparatus (see *Figures 2* and *3*) which contains the following elements:

- a cylindrical block of previously treated hardwood (leached for three days in acetone)
- a release tube, thinly coated with Vaseline on the inside surface
- a microscopic slide thinly covered with glycerol
- a metal ball (2.1 to 4 g, depending on the object studied, see below)

Method

1. Coat the inner side of the release tube (II) with a thin layer of Vaseline.
2. Shake the fully ripened fresh anther(s) gently to release pollen into tube I. Refrain from overloading the wood surface with pollen. A thin cover is preferable.
3. Remove the release tube II and fix it gently, upside down above tube III.
4. Drop the steel ball from the same distance each time (5 mm above tube II).
5. Wait 60 sec, then remove the microscopic slide and determine the average number of pollen grain in each aggregate (e.g. 200 × 5 grains for each sample). Use the data to express the distribution classes (e.g. single grain, 2–5, 6–20, 21–100, >101 grains per group).

Table 1. Main methods for airborne pollen sampling in pollination studies

Method	Advantages	Disadvantages	Applications
Gravity slide: a horizontal slide covered with adhesive which collects pollen	Cheap and easy to prepare, to handle, and to extract the pollen for examination. Many slides can be used simultaneously	The collection efficiency depends on the particle size, wind speed, wind direction, and turbulence. Gives only an indication of particles present and a rough estimation of their amount. Samples taken at various meteorological condition are not valid for comparisons	Estimation of the pollen rain in various locations in the canopy and around the plant at a given period
Rotorod sampler: The pollen is collected by standard vertical bars which are coated with adhesive and rotating at constant speed	The air volume sample is known. High efficiency of pollen collecting. Can be operated by battery in remote locations. It is commercially available	The efficiency depends on the particle size and density. Removal of the pollen from the rods is not convenient	Quantitative survey of the airborne pollen at known air volume
Rotoslide sampler: The pollen is collected by upright slides which are coated with adhesive and fixed on a rotating cross-arm at a constant speed	The air volume measured is known. High efficiency of pollen collecting. The slides are easy to insert, store, and examine. It is commercially available	It may be overloaded during prolonged sampling periods. The efficiency depends on the particle size and density. It requires electricity. It is subjected to wind impaction if exposed while idle	Quantitative survey of the airborne pollen at known air volume
Burkard's recording volumetric trap: A vacuum pump collects samples for a week on an adhesive-coated transparent tape on a clock-driven drum behind the entrance orifice	It measures variation in concentration of pollen with time. Reasonable efficiency. The tape is easy to insert, handle, and count separately for each day. It is commercially available	The efficiency depends on the wind speed and the particle size and density. Electric power and external vacuum pump are required	Daily (or seasonal) quantitative survey of airborne pollen

For more details and other samplers, consult Ogden *et al.*, ref. 22, pp. 74–95, the main source for this table.

210

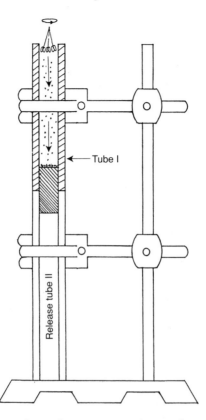

Figure 2. Stelleman's experimental set-up to produce pollen-scattering images: first stage. Tube I is clamped over the wooden block (cross-hatched) held in the release tube II, also firmly clamped in place. Pollen is released at the open end of the top of tube I and is collected on the wooden block. NB: The tubes need not be of glass; Perspex is even better. Tube I can be cut from (white) plastic tubing used for insulation of electrical wiring. The tubes are drawn as if they have the same diameter, but the release tube (II) should be a little wider than tube I for easier insertion of the wooden block, and tube III (*Figure 3*) in turn wider than II for the same reason.

This method [originally developed by Knoll (27)] is especially useful when comparing different individual plants or species. The pollen scattering pattern may vary depending on the weight of the steel ball, the distance of its release, and the apparatus size. Thus the results do not represent the natural release aggregation, but the method is still useful for comparisons especially in species which are pollinated both by insects and wind at various levels. A change of ball weight or its falling distance may produce different sizes of pollen clumps. Mohr (17) calibrated the apparatus by a comparison with the pollen 'scattering image' obtained by pollen traps (made from microscopic

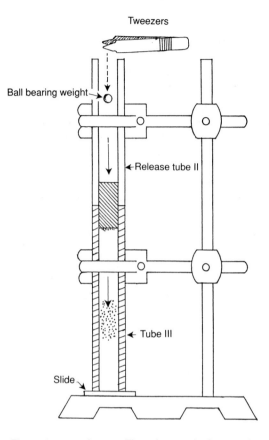

Figure 3. As *Figure 2*: second stage. The release tube is turned upside down and fits into tube III. A ball-bearing weight is dropped down on to the wooden block, releasing pollen adhering to the (now) lower surface, and caught on a glass slide for microscopical examination. (*Figures 2* and *3* adapted from ref. 13, courtesy of A. D. J. Meeuse.)

slides smeared with a sticky medium, placed at various distances from a natural pollen source). It is clear, especially for demonstration purposes, that strictly wind-pollinated species are transported mainly as single grains, while animal-pollinated pollen are clumped into aggregates. Ambophilous species (pollinated by both agents) give an intermediate picture. It should be noted that fresh pollen may be somewhat more sticky than dry pollen (26), so that fresh pollen has to be used, or the method can be employed to compare the stickiness of pollen at various ages, since in some species the pollen-kit dries quickly after pollen exposure toward a shift from insects to wind pollination (Dafni, unpublished).

2. Water pollination

Water pollination (hydrophily) is a rather rare phenomenon limited to few families and genera most of which are monocotyledons (28–30). True hydrophily (see below 'hypohydrophily') occurs in 18 submersed genera; of these 17 are monocots and 12 are marine (31). It is quite surprising that most aquatic angiosperms are insect-pollinated and wind-pollinated, while the marine ones are solely hydrophilous (32).

Hydrophilous pollination involves the use of water as a vector in the transportation of pollen. It is not necessary for the pollen itself to come in contact with water because water-borne conveyances such as anthers or flowers may carry dry pollen across the water surface (33).

Water-pollinated species tend to share some common characteristics (ref. 5, p. 285; and refs 33 and 34) such as reduced perianth, usually only one ovule per flower, commonly unisexual flowers, large and rigid stigmas, reduction in the exine of the pollen grain and elongation of spherical pollen grains.

Hydrophilous plants can be divided into three pollination categories (28, 29, 33, 35; *Table 2*).

Epihydrophily—two-dimensional pollination at the water surface, it includes two types: (a) Wet-epihydrophily in which the pollen floats just below the water surface in direct contact with the water; and (b) Dry-epihydrophily in which the reproductive structures are carried above water level and remain dry.

Hypohydrophily—tridimensional pollination below the water surface.

It is clear that wet-epihydrophily and hypohydrophily require greater modification and adaptation of reproductive structures than does dry-epihydrophily. Many of these modifications no doubt reflect specialization to facilitate pollen release and capture in water, a medium that, in addition to being wet, is significantly more viscous than air (35, 36).

Hydrophily requires the abandonment or modification of a suite of characters that are intimately tied to the dry, aerial flowering conditions. The mechanical and biochemical ramifications entailed in adapting to underwater release, transport, and capture of pollen, raise significant adaptive obstacles (37).

The fact that flowers of hydrophiles are wet when anthesis occurs makes them unique (33–35).

Observations on hydrophilous plants can be made in aquariums or special containers devised for biological studies (38) while study under natural conditions needs diving with snorkel gear (39). The method of collecting data on flower morphology, phenology, or sex distribution does not deviate from those of terrestrial plants. Observations on pollen dispersal below or above

Table 2. Main characteristics of hydrophilous flowers

Pollination category	Location of anthesis	Dimensions of pollen transfer	Reproductive structures at anthesis	Stigma	Pollen
Hypohydrophily	Below water surface	Tridimensional	Wet	Stiff, occasionally pappillate. Wettable	Filamentous or spherical. Wettable
Epihydrophily: Wet	On the water surface	Two-dimensional	Wet	Rigid, filiform Wettable	Elongate to filamentous Hydrophobic
Dry	Above water surface	Two-dimensional	Dry	Floating on the surface. Unwettable	Spherical Dry

Sources: refs 30, 35, 36, 37

the water surface may involve direct observations (40) or the use of dye or small particles (41, 42).

Anthesis, anther dehiscence, and pollen release were observed underwater with submerged light (41). In principle, dry-epihydrophylous plants could be approached and studied by the same regular methods of land plants, but the water medium may dictate the use of floating rafts; e.g. in caging experiments (43).

References

1. Faegri, K. and van der Pijl, L. (1979). *The Principles of Pollination Ecology*. (3rd edn.). Pergamon, Oxford.
2. Regal, P. J. (1982). *Annu. Rev. Ecol. Syst.*, **13**, 497.
3. Whitehead, D. R. (1969). *Evolution*, **23**, 28.
4. Whitehead, D. R. (1983). In *Pollination Biology*. (ed. L. Real), p. 97. Academic Press, Orlando, Florida.
5. Proctor, M. C. F. and Yeo, P. (1973). *The Pollination of Flowers*. Collins, London.
6. Crane, P. R. (1986). In *Pollen and Spores*. (ed. S. Blackmore and I. K. Ferguson), p. 179. Symposium Series 12. Academic Press, London.
7. Dowding, P. R. (1987). *Int. Rev. Cytol.*, **107**, 421.
8. Tauber, H. (1965). *Dan. Geol. Unders.*, (Afh.), Racke 2 No. 89, pp. 1–70.
9. Tauber, H. (1977). *Bot. Ark.*, **32**, 1.
10. Niklas, K. J. (1985). *Bot. Rev.*, **51**, 328.
11. Niklas, J. K., Buchman, S. L. and Kerchner, V. (1986). *Am. J. Bot.*, **73**, 966.
12. Niklas, J. K., (1987). *Am. J. Bot.*, **74**, 74.
13. Stelleman, P. (1982). De betekenis van de biotische bestuiving big *Plantago lanceolata*. PhD. Thesis, Department of Botany, University of Amsterdam.
14. Bino, R. J., Dafni, A., and Meeuse, A. D. J. (1984). *Proc. Ned. Kon. Aka. Wet.*, ser. C, **87**, 1.
15. Sacchi, C. F. and Prince, P. V. (1988). *Am. J. Bot.*, **75**, 1387.
16. Meeuse, A. D. J. (1990). *Flowers and Fossils*. Eburn, Delft.
17. Mohr, O. W. P. (1986). *Ephedra*—an entomophilous gymnosperm. MSc. project, Department of Botany, University of Amsterdam (Mimeo).
18. Dafni, A. and Dukas, R. (1986). *Pl. Syst. Evol.*, **154**, 1.
19. Cox, P. A. and Knox, R. B. (1988). *Ann. Missouri Bot. Gard.*, **75**, 811.
20. Scariot, A. D., Lleras, E. and Hay, J. D. (1991). *Biotropica*, **23**, 12.
21. Bawa, S. K. and Crisp, J. E. (1980). *Ecology*, **68**, 871.
22. Ogden, E. C., Raynor, G. S., Hayes, J. V., Lewis, D. M., and Haynes, J. H. (1974). *Manual for Sampling Airborne Pollen*. Hafner Press, New York.
23. Faegri, K. and Iversen, J. (1989). *Textbook of Pollen Analysis* (4th edn) (ed. K. Faegri, P. E. Kaland, and K. Krzywinski). John Wiley, London.
24. Käpylä, M. (1989). *Grana*, **28**, 215.
25. Heard, T. A., Vithanage, V., and Chacko, E. K. (1990). *Aust. J. Agric. Res.*, **41**, 1101.

26. Meeuse, A. D. J., de Meijer, A. H., Mohr, O. W. P., and Welliga, S. M. (1990). In *Flowers and Fossils* (ed. A. D. J. Meeuse) (Suppl. pp. 1–9). Eburn, Delft.
27. Knoll, F. (1930) *Flora* **166**, 43.
28. McConchie, C. A. (1982). In *Pollination '82* (ed. E. G. Williams, R. B. Knox, J. H. Gilbert, and P. Bernhardt), p. 148. School of Botany, University of Melbourne, Parkville.
29. Cook, C. D. K. (1982). In *Studies on Aquatic Vascular Plants.* (ed. J. J. Symoens, S. S. Hooper, and P. Compére), p.1. Royal Botanical Society of Belgium, Brussels.
30. Philbrick, C. T. (1991), *Rhodora*, **93**, 36.
31. Less, D. H. (1988). *Ann. Missouri Bot. Gard.*, **75**, 819.
32. Cook, C. D. K. (1988). *Ann. Missouri Bot. Gard.*, **75**, 768.
33. Cox, P. A. (1988). *Annu. Rev. Ecol. Syst.*, **19**, 261.
34. Pettitt, J. M. (1984). *Mar. Biol. Ann. Rev*, **22**, 215.
35. Philbrick, C. T. (1988). *Ann. Missouri Bot. Gard.*, **75**, 836.
36. Ducker, S. C. and Knox, R. B. (1976). *Nature (Lond.)*, **263**, 705.
37. Ducker, S. C., Pettitt, J. M., and Knox, R. B. (1978). *Aust. J. Bot.*, **26**, 265.
38. Vogel, S. and LaBarbera, M. (1978). *Bioscience*, **28**, 638.
39. Cox, P. A. (1991). *Biotropica*, **23**, 159.
40. Cox, P. A., Elmquist, T., and Tomlinson, B. P. (1990). *Biotropica*, **22**, 259.
41. Ackerman, J. D. (1983). *Biol. Bull.*, **165**, 504.
42. Ackerman, J. D. (1986). *Aquat. Bot.*, **24**, 343.
43. Osborn, J. M. and Schneider, E. L. (1988). *Ann. Missouri Bot. Gard.*, **75**, 778.
44. Di-Giovanni, F. and Kevan, P. G. (1991). *Can. J. For. Res.* **21**, 1155.

A1

Trapping and marking of foragers

1. Trapping and marking of insects

1.1 Pollen load on the insects

There is a need to distinguish between the pollen load on the various parts of the body of pollen carriers and the pollen on those parts which may be in direct contact with the receptive stigma while staying or visiting the flower. A total wash of the pollen grain gives an estimation of the 'total pollen load' (TPL) that differs from the 'functional pollen load' (FPL)—the pollen that has the chance, due to its localization on the animal body, to reach the target stigma. The chances for pollination depend on the functional pollen load, while the size and localization of the total pollen load may cause pollen waste or improper pollen transfer and clogging of the stigmas when interspecific visits are carried out.

While studying the pollen load on the forager, it is important to check the pollen quantity and identity (e.g. as indirect evidence for interspecific visits), and to map it on the pollinator body. In some orchids, an exact localization of the pollinaria on the insect body enables several orchid species to use the same pollinator without any interference (1).

The exact localization of the pollen on the agent, in relation to the spatial position of the stigma, may frequently show that many visitors, although loaded with the proper pollen, are not potential pollinators but just thieves.

Localization of the pollen on the insect body could be done by a dissecting microscope with a lower magnification than that which is needed for pollen identification. The use of SEM (2, 3) enables exact identification and quantitative localization of the pollen grains without damaging of the object. The insects should be kept in separate vials after the capture and soon after demobilized (deep freeze, CO_2, ethyl-acetate) to prevent loss or re-distribution of the pollen on the insect body. By detailed examination of the pollen distribution on various parts of the insect body, a 'pollen map' can be achieved. Clear cellotape (or adhesive tape, e.g. Scotch®) can be used to remove pollen from different parts of the animal body (insects as well as others). The clear piece of cellotape (2–4 cm) is placed directly on to a microscope slide for further identification in the laboratory or for acetolysis (Chapter 3, *Protocol 12*).

This method is also easily applicable to vertebrates. Samples from different parts of the body will produce the pollen distribution map on the animal. This method represents only partial pictures of the animal total pollen load, and the results largely depend on which body parts were sampled.

Protocol 1. Examination of the pollen load on the insect

Materials

- ethanol 95%
- gelatine–fuchsin or polyvinyl–alcohol mountant or modified Calberla's solution

Method

1. Capture the insect and immobilize it in a separate vial containing a piece of blotting paper with ethyl acetate (or another killing agent).
2. Pin the insect on to an entomological pin and wash it with small droplets of absolute ethanol.
3. Carefully collect the dripping alcohol which washes the insect body. Try to collect as much of the liquid as possible on to a microscopic slide. The insect may be squeezed gently without being harmed. Keep the insect, with all details listed (location, date, flower species, etc.), for further identification and preservation as a voucher specimen, etc.
4. Let the alcohol evaporate and repeat the procedure gently, taking care not to smear the liquid on a large area of the collecting slide.
5. Apply melted gelatine–fuchsin (Chapter 3, *Protocol 8*) or polyvinyl–alcohol mountant (chapter 3, *Protocol 10*) or modified Calberla's solution (Chapter 3, *Protocol 9*) and mark the preparation with the same number as the insect.
6. Examine the slide for pollen identification, counting, etc.

The use of individual vials is essential to prevent contamination by pollen grains in a common killing-jar. If the pollen load on the individual insect is too large to be examined in one sample, dilute the sample as in the pollen-counting procedure (Chapter 2, *Protocol 3*).

By this procedure, most of the pollen carried on the insects body is washed off; however, some still remains. Be careful to apply the ethanol to different parts of the insect body in such a way that is will be washed from different angles.

Pollen may also be removed from insects by sonification (4). The insects are caught and preserved in acetone; later they are treated with an ultrasonic probe and checked by microscope for complete removal (see Chapter 3, *Protocol 11*). Corvicula pollen pellets maintain their cohesion in acetone and have to be removed before processing.

The pollen sample is evidence of what was carried by the particular insect *but* not of the location of where it was carried on its body. Localization of pollen on the forager is sometimes essential to evaluating its role in pollination. The presence of specific pollen is evidence of a visit in a particular flower species, but does not necessarily mean that the given insect is also a 'legitimate' pollinator. Keeping a reference collection of the washed insects (which remain intact if treated gently) is essential in many studies, especially if the entomological fauna is poorly known.

The gelatine method is also applied to vertebrates (5). Small blocks (2 mm^3) of glycerol-jelly [gelatine–fuchsin (Chapter 3, *Protocol 8*) has the advantage of instant dying and a long preservation capacity] are rubbed on various areas of the animal body and the pollen grains adhere to them. Then the blocks are transferred to microscopic slides, warmed gently, and checked for pollen presence. The authors noted that the way in which the block is rubbed on the animal should be standardized. The number of the pollen grains on a slide depends upon where and how the gel block is rubbed over the animal, its size and stickiness, and the way that the animal has been handled.

Protocol 2. Removal of the total pollen load on the insect (6)

Materials

- Ethanol 70%
- glycerol
- rotator

Method

1. Capture the insect.
2. Place it in a pollen-free clean individual vial.
3. Add ethanol 70% or a mixture of 70% ethanol and glycerol (4:1) to cover the insect completely.
4. Put it in a rotator for 24–48 h.
5. Centrifugate the liquid and remove the supernatant.
6. Use the rest of the pollen for acetolysis (Chapter 3, *Protocol 12*) or transfer to gelatine–fuchsin (Chapter 3, *Protocol 8*).
7. The insect can be used now for gut-content and/or fecal analysis by careful dissection without the problem of pollen contamination from the external load.

1.2 Insect marking

Insects (or any other foragers) are marked for the following purposes:

- to follow the behavioural patterns and the foraging bouts of the individual forager
- to estimate the forager population size and intra- and interplant movements
- to determine the sharing of pollinators by plant species
- to determine the forager travel distance, the rate of dispersal, and longevity

Since the recapture rate of marked insects may be very low (1–5%, ref. 7, p. 73), a large number of insects (100–500) are marked so as to have a good-sized sample. The choice between individual and mass marking depends on the study purposes and on the advantages and disadvantages of each method (*Table 1*).

Protocol 3. Individual marking of insects

Materials

- numbered coloured discs and glue
- anaesthetic agents (carbon dioxide or ethyl acetate) or a marking device (see below)

Method

1. Catch the desired insect.
2. Immobilize it with an anaesthetic or with the special device.
3. Apply the numbered discs on the insect's thorax with the glue.
4. Release the insect as soon as it is active again.
5. Take notes on: the insect's identity, time, location, and the plant species visited, for further analysis.

Prepared numbered discs are produced by Graze (C. H. R. Graze, K. G. Fabrik für Bienegeräte, 7057 Endershach bei Stuttgart, Postfach 7, Würtemberg, Germany. Article No. 1373). Each disc is numbered from 1 to 99, and five colours are supplied (yellow, red, white, blue, and green) in each package, which limits its use to 500 marked individuals. Each package includes a bottle of glue. In principle, any adhesive label which can be easily applied to the insect without harm is suitable, depending on how conspicuous and durable it is under field conditions.

A simple marking method (especially for non-hairy insects; is to apply a

Table 1. Main methods for marking insects in pollination studies

Method	Advantages	Disadvantages
Individual marking by: numbered tags or labels	Individual identification of the forager. Easy recognition	Immobilization is needed to ensure glue hardening. Using self-adhesive labels or tapes may solve this problem
Colour spot code with quick-drying paint	Cheap and easy to carry out. Also suitable for hairy insects	Identification is not instant, the code needs to be used
Marking with typing correction fluid	Cheap and easy to carry out Suitable especially for non-hairy insects	Not for individual marking unless a number is added
Individual ferrous metal tags	Individual identification of bees	Useful only for honeybees which return to the hive in which the recapture device is located
Mass marking: with fluorescent powder, powder dye, or with any coloured dust	Cheap and easy to carry out under field conditions	The dust may interrupt the natural behaviour of the insect, especialy if attached to its compound eyes. In general, dusts are not very durable, especially on smooth and hairless insects (ref. 7, p. 74) and in wet weather. Fluorescent powder is best detected at night and the range of the common UV lamp is limited (2–10 m)
With radioactive isotopes	Applicable especially in studies of honey-bee nutrition, and rates of metabolism	Only suitable for social insects. Needs facilities to monitor radioactivity. Most impractical for field behavioural studies

white patch on the thorax with typing correction fluid (e.g. Tipp-Ex®), and when dry to add a number with a very fine technical pen (ref. 8, p. 190). The use of numbered discs is very convenient, especially for medium to large insects, and if they are not too hairy.

Any paint that is used to mark insects must be easy to apply, and to dry quickly; to be clearly distinguishable from a distance, and from which colour combinations can be obtained for the identification of individuals. A large variety of paints and dye mixtures are used to mark insects; for example, enamel paint, artist's oil paint, inks, melted crayons, lacquers, nail-varnish, mixed powders with glue, etc. (ref. 7, pp. 74–5; ref. 9, pp. 59–62). The paints are applied with fine brushes, entomological pins, toothpicks, etc., according to the type of marking agent and the specific insect. The simultaneous use of two or three colours permits the production of a number code for individual identification (ref. 10, p. 14; ref. 11).

The anaesthetics in carbon dioxide or in ethyl acetate do not harm the insects while ether, hydrocyanic acid vapours, chloroform or nitrogen and nitrous oxide may produce abnormal behaviour or even shorten the lives of the insects (ref. 7; p. 73; ref. 9, p. 74).

Keeping the insects for several minutes at near freezing temperatures (1–5 °C) immobilizes them for the sufficient time needed for marking. An insect-holding device (Graze article No. 1373) is composed of a plastic cylinder (diam. 30 mm, 7 cm long); which is covered on one side by a coarse plastic net (4 × 5 mm mesh). The cylinder is equipped with a suitable soft piston. The insect caught is placed inside the cylinder and is pushed gently against the net until it cannot move freely. The numbered disc is then applied through the hole in the net, and the insect is released. The device can easily be prepared, using a large plastic syringe in which the bottom is removed and replaced by a coarser flexible net.

Another simple technique is to dust the individual forager with fluorescent powder while visiting in flowers and later to trace the train of powder with an ultraviolet lamp (ref. 9, p. 61). The disturbance may change the behavioural pattern of the individual, but at least the later-visited flower can be detected without constant watching, but the information of the flower-visiting sequence is lost.

Protocol 4. Mass marking of insect with fluorescent powder (12)

Materials

- fluorescent dye
- a plastic bucket
- a metal wire screen (see below)
- a bicycle air-pump

Method

1. Catch a sample of the desired insects with a net.
2. Put a layer of fluorescent powder in the bottom of a plastic bucket which is covered with a metal net.
3. Put the net above a metal wire screen and blow the fluorescent dye powder by using a bicycle air-pump.
4. Release the insects.
5. At recapture follow the marked bees with an ultraviolet lamp (better at night).

Using this method, the insects are not anaesthetized, and large samples of insects can be marked in a short time and under field conditions. According to Frankie (12), the death rates of marked bees are no different from those of non-marked ones, and the marking lasts at least for six days.

Mass marking (or of individuals) with fluorescent powder enables one to detect which flowers were visited by a specific species of foragers and to detect the sharing of pollinators. Various dye colours for various species of foragers can be used simultaneously to follow the repertoire of the visited plant species by the various foragers. Marking insects at specific sites with different colours may also give an estimation of the exchange of foragers—for example, among flowering trees (12).

A large number of bees can be marked by spraying with a hand atomizer. The paints included are composed of titanium oxide in alcohol (the result is a white colour), basic fuchsin (or any other powder dye), or a fluorescent powder that can be detected by catching the foragers and viewing them under UV light (ref. 3, p. 75).

With social bees every individual returns regularly to the beehive, thus permitting the marking of individuals while entering the beehive, or by recapturing them individually on arrival. It is possible to mark all the bees in the hive by having them pass over a special hive-entrance block coated with a fluorescent powder (13). Bees from a hive may also be marked by feeding them with radioactive isotopes (ref. 9, pp. 63–71).

Gary (14) marked individual bees with ferrous metal tags in the field. Each tag has a tiny numbered plastic disk (by Graze). The bees were released after the capture and marking in the field, and the tags with the discs were later recovered by means of magnets installed at hive entrances. The recapture rate is high (60–90%) which renders this method very valuable for honey-bees (15).

Individual electronic tags with automatic recording [T. Allen (16) and in Heinrich, (ref. 17, pp. 113–114)] through an electronic tagging system, can be used to keep track of the foraging activities of individual bumble-bees, to obtain detailed information on the bees' time and energy budget, and to study patterns of behaviour in relation to food resources inside and outside the hive.

Several precautions should be kept in mind while marking insects, to minimize interference in animal behaviour. It is essential that the type of anaesthetic used does not injure the insect, inhibit its normal behaviour, or influence its life-span (ref. 7, p. 73).

The marking device should not harm the insect or disturb its natural behaviour. Comparison of the behavioural patterns between marked and non-marked individuals is desirable to clarify this issue (18).

1.3 Insect trapping

Every pollination study which involves insect identification, observation, or marking, involves trapping. The main methods applicable in pollination research are compared in *Table 2*. Technical data on trapping means and their structure and construction are available; for example, in Southwood (ref. 9, pp. 191–216), Peterson (ref. 7, p. 83, and plates 63, 64, 74, 163, 166, and 168),

Table 2. Main methods for trapping insects in pollination studies

Method	Advantages	Disadvantages
Air-flight Malaise trap	Suitable for massive day and night, collection, especially of Diptera and Hymenoptera in woodlands (ref. 9, p. 163). Easy to use, low cost. May success-fully be used with massive marking in close habitats. The trapped insects are alive and can be used for marking and experimentation	Not practical under windy situations. Unsatisfactory for Coleoptera and Hemiptera (ref. 9, p. 193). The wind direction may considerably influence the trapped yield; any change of position or location may also influence the composition of the trapped insects (ref. 20, p. 155)
Window trap	Very efficient for trapping beetles and bees in open habitats	The water in the collection tray may wash off some of the pollen load. The trapped insects have to be removed immedi-ately if they are needed alive
Sticky traps	Efficient especially for mosquitoes, thrips, small Coleoptra, Hymenoptra, and Diptera	There is a need to dissolve the sticky agent, a procedure which may wash off some of the pollen from the insect. The trapped insects are dead and cannot be used for further observations. Large insects (e.g. honey-bees) may escape from the trap and will thus be underestimated. Pollen sample from the sticky surface could also be analysed (22)
Water traps	Very efficient if the pollinator has a colour preference. The traps can be left, and the yield can be gathered at any time if there is no rain	The pollen load on the insect is partially washed off. The trapped insects have to be removed immediately if they are needed alive
The Robinson trap	Gives the maximum catch of larger Lepidoptera. Insects may be caught alive or killed. Of all the many types, it seems that this is especially practical for hawk-moths (ref. 9, p. 204)	The results may not reflect the exact relative abundance of different species. The trap is not protected from rain although it has a drainage funnel in its bottom

Oldroyd (ref. 19, pp. 26–62), and Muirhead-Thompson (20). The habitats, the forager type and/or its behaviour should determine which specific trap should be utilized.

The main traps relevant to pollination studies are:

The Malaise trap
The Malaise trap consists basically of an open-fronted tent of cotton or plastic net, black or green in colour. The 'roof' slopes upwards to the innermost corner at which there is an aperture leading to a trap. For exact technical details, see Muirhead-Thompson (ref. 20, pp. 153–7).

The window trap
This trap consists of a vertical clear glass panel, below which is a trough containing water with some detergent and a little preservative (21).

The sticky trap
Sticky traps are composed of an adhesive surface to which the insects stick when they land or tangle with while flying. The adhesive should be purchased commercially ('Tangle-trap insect trapping adhesive' by Tanglefoot Company, USA) or be prepared from grease or resin. House (21) used wire screens (galvanized mesh size 12.5 × 12.5 mm, wire diameter of 0.8 mm, screen size 400 × 250 mm) which were painted with insect-trapping adhesive. The screens were fixed with a curtain that was attached to the upper and lower mid-point of the screen frame, with a nylon cord tied to each ring. At the end of the sampling period each screen was individually packed to prevent contamination and later soaked in a kerosene bath in the laboratory for 10–20 min. The insects were then removed individually and placed in 70% alcohol for identification and to check the pollen load. (The kerosene was collected and filtered later for pollen analysis.) For other methods consult Southwood (ref. 9, pp. 197–8) and Muirhead-Thompson (ref. 20, pp. 181–7).

The Robinson trap
This trap has no roof; the UV light source is protected by a clear Perspex cone. The bulb is situated in the middle of the funnel which is equipped with slides leading to the catching drum. A killing agent may be placed inside the drum (ref. 19, pp. 49–50; ref. 9, p. 204). For other types of light traps see Muirhead-Thomspon (ref. 20, pp. 4–11).

Water traps
These traps are just containers filled with water and some detergent. The traps may be transparent or painted in various colours and placed at any height. If the pollinator shows a preference for a certain colour (e.g. *Amphicoma* spp. beetles; ref. 23) then water traps are very efficient and can be used as flower models of various shapes and sizes (Dafni, unpublished). Using traps of various colours may attract selectively different groups on insects (24).

2. Trapping and marking of birds (by P. Feinsinger)

2.1 Pollen load on the birds

The first consideration in dealing with avian flower-visitors is that two fundamentally different groups exist: humming-birds (family Trochilidae), restricted to the New World, and, on all continents but Europe, a variety of birds from other families, chiefly but not exclusively in the order Passeriformes. Not only do humming-birds differ from other birds in their mode of flight and nectar preferences (25, 26), but also their unique morphologies dictate unique techniques for removal of pollen loads and for marking individuals.

Most pollen that plants deposit on humming-birds is carried on the exposed (unfeathered) part of the beak, or on the feathered forehead or chin. A few plant species deposit pollen on the crown or nape, fewer yet on the throat (gorget). Very little pollen ends up on the sides of the heads. Thus, two strips of cellophane tape are sufficient to sample precisely the pollen loads carried by a humming-bird, maintaining the original spatial position of each grain. One tape is carefully oriented on the dorsal surface, extending from beyond the bill-tip backwards to the base of the nape, and then is pressed down and rocked back and forth, thereby removing all grains not firmly adhering to the bill or feathers (27, 28). The tape is mounted on a microscope slide such that the impression made by the tip of the bill is positioned exactly at the edge of the slide. A second cellophane tape is used to sample the ventral surface, from bill-tip to the base of the gorget, and mounted on a separate slide. Precise measurements should be taken with a flexible tape, with origin at the bill-tip, of the following points on the dorsum: millimetres to base of unfeathered part of bill; to base of mandible (usually about 5 mm further back); to arbitrary division between forehead and crown (see 29 for details); to arbitrary division between crown and nape. Likewise, on the venter the distance from bill-tip to first feathering should be noted; next, distance to arbitrary division between chin and throat. If it is difficult to make such measurements then each sample can simply be divided into 5-mm segments, beginning with the tip.

So-called 'invisible' tapes (such as Scotch® brand 'green plaid' tape) should be avoided, as they are quite visible under the microscope; in contrast, Scotch® brand 'red plaid' tape (3M Corporation, USA) is the tape of choice when available. Other brands do not appear to last the 5–8 years typical of the latter (30).

With this technique, no staining of grains is possible, but with proper optical equipment staining is not necessary. An inexpensive laboratory microscope that has been modified with Hoffman® Modulation Optics on the 20× objective makes visible the three-dimensional surface texture of pollen grains and facilitates identification enormously. Better still is a high-quality microscope equipped with Nomarsky differential interference contrast

(DIC) optics providing an excellent view of the three-dimensional structure of grains. Cellophane tape does not interfere noticeably with the optical quality. More sophisticated tecniques such as scanning electron microscopy (31) can be utilized with double-sided sticky tapes or similar pollen removal methods that leave the mounted grains exposed, but such techniques require considerable effort in order to quantify and map the entire pollen load carried by a bird.

An alternative technique for removing pollen from humming-birds is the 'acid fuchsin' technique of glycerol-jelly squares (Chapter 3, *Protocol 8*). As several squares are needed just to sample the bill, however, and as the exact placement of grains within each square is lost, the cellophane-tape technique is typically more efficient, unless the investigator has a strong reason for using the gelatine-square technique. On the other hand, the gelatine-square technique is often preferable for sampling pollen loads on other avian flower-visitors, which often carry much more pollen on the head feathers than on the bill (cf. refs. 32, 33, and 34). A simpler technique that may be no less precise simply consists of combing each successive small patch of feathers on the bird's head, using a fine mascara brush, on to a microscope slide coated with a thin layer of Vaseline petroleum jelly (33). A modification of the cellophane-tape technique may also suffice, in which small pieces of tape are pressed against specified parts of the bird's plumage, then mounted each on a separate labelled slide (35).

No pollen sample taken from a bird should be viewed as a complete, quantitative measure of the pollen grains with which the bird was laden when captured. Even when handled with the utmost of care, birds will undoubtedly have lost a lesser or greater portion of their pollen load by the time the sample is taken. If the spatial distribution of pollen grains on the bird's bill and feathers is of interest to the investigator, then equally serious is the chance that pollen loads may be 'smeared' during the capture process. The standard method of capturing birds with mist nets leads to unavoidable loss of pollen and potential pollen smearing as the bird first hits the net head-on, then struggles to escape. Although normal mist-netting practices do not require one to keep constant vigilance at nets, the pollination ecologist should be prepared to run immediately to the net when a bird is captured and to remove the pollen, minimizing direct contact with the pollen-carrying body parts, before extracting the bird from the net.

2.2 Capturing the birds

Mist-netting must only be done by qualified researchers that have been certified and licensed to utilize this technique (for example, in the United States by the US Fish and Wildlife Service). In most nations it is necessary to obtain a government permit to capture birds using any technique whatsoever, and often necessary to obtain a special permit for the use of mist-nets. With proper certification, mist-nets and othe materials can be obtained from Avinet Inc. Note that nets of different mesh sizes are recommended for capturing different size classes of birds. Although 1-in mesh nets are often recommended for humming-birds these nets do not capture significantly greater numbers of humming-birds than more durable 1¼-in mesh nets. Researchers working with nectar-feeding birds other than humming-birds should use either 1¼-in or 1½-in mesh nets, the latter particularly in the Australasian region where quite large species are involved.

2.3 Marking the birds

Birds other than humming-birds are effectively marked with coloured plastic rings. Contrasting, bright colours should be used when possible, up to a maximum of two rings per leg. Because humming-birds draw their legs up into the ventral plumage when flying, and often have feathered tarsi that obscure rings even when birds are perched, coloured rings are not suitable for marking humming-birds. No other technique is without drawbacks. Stiles and Wolf (36) provide precise instructions for cutting leg-flags from coloured acetate sheets, forming the leg-flags by bending them around a dissecting needle (approximately the same diameter as a humming-bird's leg), coding the flag with small (2 mm wide) strips of coloured tape, and using Duco® Household Cement (not Duco® Plastic Cement) to bond the flag to itself after carefully wrapping it around the humming-bird's leg. If this technique is used, the researcher should take the following precautions:

(a) make sure that the colours of coding tape used are bright and easily distinguished from other possible colours (e.g. brown, black, and dark blue are virtually impossible to distinguish in forest understorey);

(b) make sure that no cement gets on the bird's leg; and

(c) make sure that the flag is not so tight as to restrict circulation, but not so loose as to slide off the foot.

The acetate-flag technique has been used successfully by many authors (see, for example, refs. 37, 38, and 39). Nevertheless, even with the most careful technique, the circulation may be restricted, or else the constant rubbing of the acetate may wound the leg and allow pathogens to invade, resulting in loss of the leg (Feinsinger, pers. obs.; cf. ref. 40). Furthermore acetate flags may interfere with nesting (41). No adequate alternative material for leg-flags has been developed; for example, soft vinyl flags do not bond well and do not retain coloured tape well (H.M. Tiebout III, pers, comm.), whereas coloured strings or streamers attached to the leg (cf. ref. 42) may possibly become entangled in vegetation. Therefore, unless it is absolutely necessary to have long-lasting marks, alternative techniques should be employed. Many researchers mark the back of humming-birds with small coded spots of enamel model-airplane pain (such as Testor's®), but care must be taken to avoid excessive damage to plumage; furthermore, it is possible that such colours might affect social status. Baltosser (40) suggests using bicoloured discs, 4 to 6.8 mm in diameter, made from plastic-coated nylon fabric. Discs are glued to the back feathers of humming-birds with Duco® Household Cement. While discs will be lost at the time of the next moult, in contrast to paint dots (which wear off with time) the discs retain their usefulness up until the moment of moult. The diversity of codes can be increased not only by using variously coloured fabric but also by using a variety of geometric shapes (40). Thus, an acceptable compromise is to mark a captured humming-bird using Baltosser's technique but simultaneously band it with a standard metal bird-band, such that, if the bird is by chance recaptured after it has lost the original disc, it can still be identified and re-marked.

References

1. Williams, N. H. (1982). In *Orchid Biology: Reviews and Perspectives* (ed. J. Arditti) p. 119. Cornell University Press, Ithaca, NY.
2. Turnock, W. J. and Chong, J. (1978). *Canad. J. Zool*, **56**, 2051.
3. Stelleman, P. (1978). *Acta Bot. Neerl.*, **27**, 333.
4. Berger, L. A., Vassiere, B. E., and Moffett, J. O. (1988). *Environ. Ent.*, **17**, 789.
5. Wooler, R. D., Russell, E. M., and Renfree, M. B. *Aust. Wildl. Res.*, **10**, 433.
6. Janzon, L. A. (1983). *Grana*, **25**, 153.
7. Peterson, A. (1953). *A Manual of Entomological Techniques*. Ewards Brothers., Ann Arbor, Michigan.
8. Heinrich, B. (1983). In *Handbook of Experimental Pollination Biology* (ed C. E. Jones and R. J. Little), p. 187. Van Nostrand Reinhold, New York.
9. Southwood, T. R. E. (1966). *Ecological Methods*. Chapman & Hill, London.
10. Frisch, von K. (1967). *The Dance Language and Orientation of Bees*. Belknap Press of Harvard University Press, Cambridge, Mass.
11. Heinrich, B. (1976). *Ecology*, **57**, 874.
12. Frankie, G. W. (1973). *Ann. Entomol. Soc. Am.*, **66**, 690.
13. Smith, M. V. and Townsend, G. F. (1951). *Canad. Entomol.*, **83**, 346.
14. Gary, N. E. (1971). *J. Econ. Ent.*, **64**, 961.
15. Gary, N. E. (1979). *Proc. IVth Int. Symp. on Pollination. Md. Agric. Exp. Spec. Misc. Publ.*, **1**, 359.
16. Heinrich, B. (1979). *Bumblebees Economics*. Harvard University Press, Cambridge, Mass.
17. Allen, T. (1981). In Heinrich (see ref. 8) in *Handbook of Experimental Pollination Biology* (ed. C. E. Jones and R. J. Little) Van Nostrand and Reinhold, New York.
18. Opp, S. B. and Prokopy, R. J. (1986). In *Insect Plant Interaction* (ed. J. R. Miller and T. A. Miller), p. 1. Springer-Verlag, New York.
19. Oldroyd, H. (1958). *Collecting, Preserving and Studying Insects*. Hutchinson, London.

20. Muirhead-Thompson, R. C. (1991). *Trap Response of Flying Insects*. Academic Press, Harcourt Brace Jovanovich, London.
21. Chapman, J. A. and Kingshorn, J. M. (1955). *Canad. Ent.*, **82**, 46.
22. House, S. M. (1989). *Aust. J. Ecol.* **14**, 77.
23. Dafni, A., Bernhardt, P., Shmida, A., Ivri, Y., Greenbaum, S., O'Toole, Ch., and L. Losito (1990). *Isr. J. Bot.*, **39**, 81.
24. Kirk, S. A. (1984). *Ecol. Ent.*, **9**, 35.
25. Baker, H. G. and Baker, I. (1983). In *Handbook of Pollination Biology* (ed. C. E. Jones and R. J. Little), p. 117. Van Nostrand Reinhold, New York.
26. Martinez del Rio, C. and Karasov, W. (1990). *Am. Nat.*, **136**, 618.
27. Feinsinger, P., Wolfe, J. A., and Swarm, L. A. (1982). *Ecology*, **63**, 494.
28. Feinsinger, P., Beach, J. H., Linhart, Y. B., Busby, W. H., and Murray, K. G. (1987). *Ecology*, **68**, 1294.
29. Baldwin, S. P., Oberholser, H. C., and Worley, L. G. (1931). *Measurements of Birds*. Cleveland Museum of Natural History, Cleveland, Ohio.
30. Feinsinger, P. (1987). *Rev. Chile. Hist. Nat.*, **60**, 285.
31. Boeke, J. D. and Ortiz-Crespo, F. I. (1978). *Science*, **201**, 167.
32. Feinsinger, P., Linhart, Y. B., Swarm, L. A., and Wolfe, J. A. (1979). *Ann. Missouri Bot. Gard.* **66**, 451.
33. Hopper, S. D. (1980). *Aust. J. Bot.* **28**, 61.
34. Hopper, S. D. (1980). *West Aust. Nat.*, **14**, 186.
35. Paton, D. C. and Turner, V. (1985). *Aust. J. Bot.*, **33**, 271.
36. Stiles, F. G. and Wolf, L. L. (1973). *Condor*, **75**, 244.
37. Feinsinger, P. (1976). *Ecol. Monog.*, **46**, 257.
38. Stiles, F. G. (1980). *Ibis*, **122**, 322.
39. Gill, F. B. (1988). *Ecology*, **69**, 1933.
40. Baltosser, W. H. (1978). *Bird-Banding*, **49**, 47.
41. Waser, N. M. and Calder, W. A. (1975). *Condor*, **77**, 361.
42. Paton, D. C. and Carpenter, F. L. (1984). *Ecology*, **65**, 1808.

Reagents and solutions

Formalin–acetic acid–alcohol FAA (ref. 1; p. 89)

  90 ml ethanol 30% (industrial quality)
  5 ml glacial acetic acid
  5 ml formalin (38–40%)

Mix all the ingredients and store; it can be kept for years.

Glycerol-jelly (ref. 2, p. 75)

  2 g gelatine
  12 cc distilled water
  11 cc glycerol
  2% phenol (1 part to each 50 parts of the glyerol-jelly)

Mix the gelatine and water, warm slightly to dissolve the gelatine faster. When dissolved, add the glycerol and the phenol. Let stand overnight. Strain through cheesecloth. For pre-staining glycerol-jelly, add a drop of saturated aqueous basic fuchsin for each 25–30 ml of the glycerol-jelly.

Iodine–potassium iodine (IKI) (ref. 1, p. 184)

  0.2 g potassium iodine
  1 g iodine
  100 ml distilled water

Dissolve the potassium iodine in as little water as possible and add the iodine. Complete with distilled water to 100 ml.

Caution: Iodine is poisonous, do not inhale vapours. Stains starch in blue to blue-black, occasionally red to purple.

Käpylä's adhesive mixture (3)

  10 g gelatine
  70 ml glycerol
  60 ml distilled water
  0.25 g phenol

Prepared as glycerol-jelly.

Lactophenol (ref. 4, p. 100)

20 cc phenol (melted)
20 cc lactic acid
40 cc glycerol
20 cc water

Melt the phenol in a water bath and then add the other ingredients.

Lactophenol–cotton blue (ref. 5, p. 47)

20 ml phenol (pure crystals)
20 ml lactic acid
40 ml glycerol
20 ml distilled water
0.05 g cotton blue (= methylene blue)

Dissolve the phenol in water by warming gently, then add the lactic acid, glycerol, and cotton blue.

Polyvinyl–alcohol (PVA) lactophenol (Irene Baker, pers. comm.)

15 g PVA (e.g. Kodak. No. 153, 9709).
100 ml distilled water
22 ml lactic acid
22 g phenol

Dissolve the PVA, with constant stirring, in the water in a beaker of water bath of 80 °C for 30 min. Then add the lactic acid and the phenol (**with caution!**), continue to stir for 2–5 min. Keep in a dark bottle.

Sudan III and IV (ref. 6, Section 6.15.2.8)

Add excess of Sudan IV to saturated solution of Sudan III in 70% ethanol.

The solution is useable for a few days. Use 70% ethanol for rinsing. The dye is stable in solution for a few weeks in a dark bottle.

References

1. Purvis, M. J. (1966). *Laboratory Techniques in Botany*. Butterworths, London.
2. Ogden, E. C., Raynor, G. S., Hayes, J. V., Lewis, D. M., and Haynes, J. H. (1974). *Manual of Sampling Airborne Pollen*. Hafner Press, New York.
3. Käpylä, M. (1989). *Grana*, **28**, 215.
4. Sass, J. E. (1951). *Botanical Microtechniques*. Iowa State College Press, Ames, Iowa.
5. Grimstone, A. V. and Skaer, R. J. (1972). *A Guidebook to Microscopic Methods*. Cambridge University Press.
6. O'Brien, T. P. and McCully, M. E. (1969). *Plant Structure and Development*. Macmillan, Toronto.

A3

Further reading

1. Books on pollination ecology and related subjects

Armstrong, J. A., Powell, J. M., and Richards, A.J. (ed.) (1982). *Pollination and Evolution*. Royal Botanic Garden, Sydney.

Barth, H. G. (1985). *Insects and Flowers*. Princeton University Press, Princeton, NJ.

Bentley, B. and Elias, T. (ed.) (1983). *The Biology of Nectaries*. Columbia University Press, New York.

Brantjes, N. B. M. and Linskens, H. F. (ed.) (1973). *Pollination and Dispersal*. Symposium to honour L. van der Pijl. Publ. Dept. Bot., University of Nijmegen.

Butler, C. G. (1958). *The World of the Honey Bee*. The New Naturalist Series. Collins, London.

Clements, F. E. and Long, F. L. (1923). *Experimental Pollination. An Outline of the Ecology and Flowers and Insects*. Carnegie Institution of Washington, Washington, DC.

Dafni, A. and Eisikowitch, D. (ed.) (1990). *Advances in Pollination Ecology*. The Weizmann Science Press of Israel, Jerusalem. (Published also as *Isr. J. Bot.*, **39**, 3–228).

Faegri, K. and Iversen, J. (1989). *Textbook of Pollen Analysis* (4th edn.) (ed. K. Faegri, P. E. Kaland, and K. Krzywinski), John Wiley, Chichester and New York.

Faegri, K. and van der Pijl (1979). *The Principles of Pollination Ecology* (3rd edn.). Pergamon Press, Oxford.

Frankel, R. and Galun, E. (1977). *Pollination Mechanism, Reproduction and Plant Breeding*. Springer-Verlag, Berlin.

Free, J. B. (1970). *Insect Pollination of Crops*. Academic Press, New York.

Frisch, K. von (1967). *The Dance Language and Orientation of Bees*. Belknap Press of Harvard University Press, Cambridge, Mass.

Frisch, K. von (1950). *Bees: Their Vision Chemical Senses and Language*. Cornell University Press, Ithaca, NY.

Futuyma, D. J. and Slatkin, M. (1983). *Co-evolution*, Sunderland, Mass.

Gilbert, L. E. (ed.) (1975). *Co-evolution of Animals and Plants*. University of Texas Press, Austin.

Gould, J. L. and Gould, C. G. (1988). *The Honey Bee*. Scientific American Library, a Division of HPHLP, New York.

Grant, K. A. (1968). *Humming-birds and Their Flowers*. Columbia University Press, New York.

Grant, V. (1963). *The Origin of Adaptation*. Columbia University Press, New York.

Grant, V. and Grant, K. (1965). *Flower Pollination of Phlox Family*. Columbia University Press, New York.

Heinrich, B. (1979). *Bumble-bee Economics*. Harvard University Press, Cambridge, Mass.

Heywood, V. H. (ed.) (1973). *Taxonomy and Ecology*, Academic Press, London.

James, W. O. and Clapham, A. R. (1935). *The Biology of Flowers*. Clarendon Press, Oxford.

Jones, S. G. (1939). *Introduction to Floral Mechanisms*. Blackie, London and Glasgow.

Jones, C. E. and Little, R. J. (ed.), (1983). *Handbook of Experimental Pollination Biology*. Van Nostrand Reinhold, New York.

Kapil, R. P. (ed.), (1986). *Pollination Biology—An Analysis*. Inter-India Publications, New Delhi.

Knuth, P. (1906–9). *Handbook of Flower Pollination*. Trans. by J. R. Ainsworth Davis, Clarendon Press, Oxford. Vol I: 1906; Vol II: 1908; Vol III: 1909.

Kugler, N. (1970). *Blütenökologie*. Fischer-Verlag, Stuttgart.

Leppik, E. E. (1977). *Floral Evolution—In Relation to Pollination Ecology*. Today and Tomorrow's Printers and Publishers, Hissar.

Lovett Doust, J. and Lovett Doust, L. (ed.) (1988). *Plant Reproductive Ecology: Patterns and Processes*. Oxford University Press, New York.

Lindauer, M. (1971). *Communication Among Social Bees*. Harvard University Press, Cambridge, Mass.

Linskens, H. F. (1964). *Pollen Physiology and Fertilization*. North-Holland, Amsterdam.

Manning, A. (1972). *An Introduction to Animal Behaviour*. Edward Arnold, London.

McGregor, S. E. M. (1976). *Insect Pollination of Cultivated Plants*. USDA Handbook No. 436. U.S. Government Printing Office, Washington, D.C.

Meeuse, B. J. D. (1961). *The Story of Pollination*. Ronald Press, London.

Meeuse, B. J. D. (1977). *Reproductive Biology of Flowering Plants*. ASUW Lecture Notes, Botany 475, 113 HUB, FK-10, University of Seattle.

Meeuse, B. J. D. and Morris, S. (1984). *The Sex Life of Flowers*. Faber & Faber, London.

Moore, P. D., Webb, J. A., and Collinson, M. E. (1991). *Pollen Analysis* (2nd edn). Blackwell Scientific Publications, Oxford.

Mulcahy, O. L., Bergamini-Mulcahy, G., and Ottaviano, E. (ed.) (1986). *Biotechnology and Ecology of Pollen*. Springer-Verlag, Berlin.

Mulcahy, D. L. and Ottaviano, E. (ed.) (1983). *Pollen: Biology and Implications for Plant Breeding*. Elsevier Bio-Medical, New York.

Müller, H. (1883). *Fertilization of Flowers*. Macmillan, London.

Percival, M. S. (1969). *Floral Biology*. Pergamon Press, Oxford.

Proctor, M. and Yeo, P. (1973). *The Pollination of Flowers* (trans. D'A. W. Thompson). The New Naturalist Series. Collins, London.

Radford, E. A., Dickison, W. C., Massey, J. R., and Bell, C. R. (1974). *Vascular Plant Systematics*. Harper & Row, New York.

Real, L. (ed.), 1983. *Pollination Biology*. Academic Press Harcourt Brace Janovich, Orlando, Florida.

Rebelo, A. G. (ed.) (1987). *A Preliminary Synthesis of Pollination Biology in the Cape Flora*. Report No. 141. South African National Scientific Programmes, Pretoria.

Richards, A. J. (ed.) (1978). *The Pollination of Flowers by Insects*. Linnean Society Symposium Series No. 6. Academic Press, London.

Richards, A. J. (1986). *Plant Breeding Systems*. George Allen & Unwin, London.

Roubik, D. W. (1989). *Ecology and Natural History of Tropical Bees*. (Chapter 2: Foraging and pollination, pp. 25–161). Cambridge University Press.

Shivanna, K. R. and Johri, B. M. (1985). *The Angiosperm Pollen: Structure and Function*. Wiley Eastern, New Delhi.

Stanley, R. G. and Linskens, H. F. (1974). *Pollen*. Springer-Verlag, Berlin.

Stebbins, G. L. (1950). *Variation and Evolution in Plants*. Columbia University Press, New York.

Thornhill, N. W. (ed.) (1991). *The Natural History of Inbreeding and Outbreeding: Theoretical and Empirical Perspectives*. Chicago University Press, Chicago.

Vogel, S. (1954). *Blütenbiologische Typen als Elementeder Sippengliederung dargestalt anhand der flora Südafrikas*. Fischer-Verlag, Jena.

Vogel, S. (1990). *The Role of Scent Glands in Pollination—On the structure and function of osmophores*. Smithsonian Institution Libraries and the National Science Foundation, Washington, DC. [Translation of Vogel, S. (1963). *Abh. Math.—Naturwiss. Kl. Akad.*, Mainz, 1962(10), 599–763.]

Weberling, F. (1989). *Morphology of Flowers and Inflorescences*. Cambridge University Press.

Willemstein, S. C. (1987). *An Evolutionary Basis for Pollination Ecology*. E.J. Brill, Leiden.

Williams, E. G., Knox, R. B., Gilbert, J. H., and Bernhardt, P. (ed.) (1982). *Pollination '82*. Plant and Cell Biology Research Center, University of Melbourne, Parkville.

Williams, E. G. and Knox, R. B. (ed.) (1984). *Pollination '84*. Plant Cell Biology Research Centre, University of Melbourne, Parkville.

Willson, M. F. (1983). *Plant Reproductive Ecology*. John Wiley, New York.

Willson, M. F. and Burnley, H. (1983). *Mate Choice in Plants: Tactics, Mechanisms and Consequences*. Princeton University Press, Princeton, NJ.

2. Key reviews and chapters on pollination

Armstrong, J. A. (1979). Biotic pollination mechanisms in Australian flora—a review. *NZ J. Bot.*, **17**, 467–508.

Baker, H. G. (1973). Evolutionary relationships between flowering plants and animals in American and African forests. In *Tropical Forest Ecosystems in Africa and South America: A Comparative Review*. (ed. B. J. Meggers, E. S. Ayensu, and W. D. Duckworth), pp. 145–59. Smithsonian Institute Press, Washington, DC.

Baker, H. G. (1983). An outline of the history of anthecology, or pollination biology. In *Pollination Biology* (ed. L. Real). pp. 7–28. Academic Press, Orlando, Florida.

Baker, H. G. 1985. Trends in pollination ecology. *Aliso* **11**, 213–29.

Baker, H. G. and Baker, I. (1983). A brief historical review of the chemistry of floral nectar. In *The Biology of Nectaries* (ed. B. Bentley and T. Elias), pp. 126–52. Columbia University Press, New York.

Baker, H. G. and Baker, I. (1983). Floral nectar sugar constituents in relation to pollinator type. In *Handbook of Experimental Pollination Biology* (ed. C. E. Jones and R. J. Little), pp. 117–141. Van Nostrand Reinhold, New York.

Baker, H. G. and Hurd, P. D., Jr. (1968). Intrafloral ecology. *Ann. Rev. Entomol.*, **13**, 385–414.

Barrett, S. C. H. (1988). The evolution, maintenance, and loss of self-incompatibility systems. In *Plant Reproductive Strategies* (ed. J. Lovett-Doust and L. Lovett-Doust), pp. 98–124. Oxford University Press, New York.

Barrett, S. C. H. and Eckert, C. G. (1990). Variation and evolution of the mating systems in seed plants. In *Biological Approaches and Evolutionary Trends in Plants*, (ed. S. Kawano). pp. 229–54. Academic Press, London.

Bawa, K. S. (1983). Patterns of flowering in tropical plants. In *Handbook of Experimental Pollination Biology* (ed. C. E. Jones and R. J. Little), (pp. 394–410. Van Nostrand Reinhold, New York.

Bawa, S. K. (1990). Plant–pollinator interactions in tropical rain forests. *Annu. Rev. Ecol. Syst.*, **21**, 399–422.

Bawa, S. K. and Beach, J. H. (1981). Evolution of sexual systems in flowering plants. *Ann. Missouri Bot. Gard.*, **68**, 254–74.

Bell, G. (1985). On the function of flowers. *Proc. R. Soc. London, Ser. B, Biol. Sci.* **224**, 223–65.

Bertsch, A. (1987). Flowers as food sources and the cost of outcrossing. In *Ecological Studies*, Vol. 61, pp. 277–93 (ed. E. D. Schulze and H. Zwölfer). Springer-Verlag, Berlin.

Bertin, R. I. (1988). Paternity in plants. In *Plant Reproductive Strategies* (ed. J. Lovett-Doust and L. Lovett-Doust). pp. 30–59. Oxford University Press.

Bertin, R. (1989). Pollination biology. In *Plant–animal Interactions.* (ed. W. G. Abrahamson), pp. 23–86. McGraw-Hill, New York.

Borg-Karlson, A. K. (1990). Chemical and ethologial studies of pollination in the genus *Ophrys*. *Phytochemistry (Oxf.)* **29**, 1359–87.

Buchmann, S. L. (1983). Buzz pollination in angiosperms. In *Handbook of Experimental Pollination Biology* (ed. C. E. Jones and R. J. Little), pp. 73–116. Van Nostrand Reinhold, New York.

Buchmann, S. L. (1987). The ecology of oil flowers and their bees. *Annu. Rev. Ecol. Syst.*, **18**, 343–70.

Charlesworth, D. (1987). Self-incompatibility systems in the flowering plants. *Persp. Biol. Med.*, **30**, 263–77.

Charlesworth, D. and Charlesworth, B. (1987). Inbreeding depression and its evolutionary consequences. *Annu. Rev. Ecol. Syst.*, **18**, 237–68.

Corbet, S. (1990). Pollination and weather. *Isr. J. Bot.*, **39**, 13–30.

Couvillon, P. A. and Bitterman. M. E. (1991). How honey bees make choices. In *The Behaviour and Physiology of Bees.* (ed. L. J. Goodman and R. C. Fisher), pp. 116–30. C.A.B. International, London.

Cox, P. A. (1988). Hydrophylous pollination. *Annu. Rev. Ecol. Syst.*, **19**, 261–80.

Cox, P. A. (1991). Abiotic pollination: an evolutionary escape for animal-pollinated angiosperms. *Phil. Trans. R. Soc. Lond. B* **333**, 317–24.

Crane, P. R. (1986). Form and function in wind dispersed pollen. In *Pollen and Spores* (ed. W. Blackmore and I. K. Ferguson), pp. 179–200. Symposium Series. Academic Press, London, 12.

Crepet, W. L. (1983). The role of insect pollination in the evolution of the angiosperms. In *Pollination Biology* (ed. L. Real), pp. 31–50. Academic Press, Orlando, Florida.

Cruden, R. W., Hermann, S. M., and Peterson, S. (1983). Patterns of nectar production and plant-pollinators coevolution. In *The Biology of Nectarics* (ed. B. Bentley and T. Elias), pp. 80–125. Columbia University Press, New York.

Cruden, R. W. and Lyon, D. L. (1989). Facultative xenogamy: examination of mixed mating systems. *The Evolutionary Ecology of Plants* (e.d. J. H. Bock and Y. B. Linhart), pp. 171–208. Westview Press, Boulder, Colorado.

Dafni, A. (1984). Mimicry and deception in pollination. *Annu. Rev. Ecol. Syst.*, **15**, 259–78.

Dafni, A. and Bernhardt, P. (1990). Pollination of terrestrial orchids of Southern Australia and the Mediterranean region, systematic, ecological and evolutionary implications. *Evol. Biol.*, **24**, 193–252.

Dobson, H. E. M. (1989). Pollenkit in plant reproduction. In *The Evolutionary Ecology of Plants* (ed. J. Bock and Y. Linhart), pp. 227–46. Westview, Boulder, Colorado.

Dowding, P. (1987). Wind pollination mechanisms and aerobiology. *Int. Rev. Cyt.*, **107**, 421–37.

Dickinson, H. G. (1990). Self-incompatibility in flowering plants. *Bio Essays*, **12**, 155–61.

Ducker, S. C. and Knox, R. B. (1985). Pollen and pollination: a historical review. *Taxon* **34**, 401–19.

Feinsinger, P. (1983). Coevolution and pollination. In *Coevolution* (ed. D. J. Futuyma and M. Slatkin), pp. 282–310. Sinauer Associates, Sunderland, Mass.

Feinsinger, P. (1987). Approaches to nectarivore–plant interaction in the New World. *Rev. Chile. Hist. Nat.*, **60**, 285–319.

Friis, E. M. and Endress, P. K. (1990). Origin and evoluition of Angiosperm flowers. *Adv. Bot. Res.* **17**, 100–62.

Ganders, F. R. (1979). The biology of heterostyly. *NZ J. Bot.*, **17**, 607–35.

Gottsberger, G. (1988). The reproductive biology of primitive angiosperms. *Taxon*, **37**, 630–43.

Gottsberger, G. (1989). Floral ecology—report on the years 1985 (1984) to 1988. *Prog. Bot.* **50**, 352–79.

Gould, J. L. (1991). The ecology of honeybee learning. In *The Behaviour and Physiology of Bees*, (ed. L. J. Goodman and R. C. Fisher), pp. 306–22. CAB International, London.

Handel, S. N. (1983). Pollination ecology, plant population structure, and gene flow. In *Pollination Biology* (ed. L. Real), pp. 163–212. Academic Press, Orlando, Florida.

Harborne, J. B. (1982). *Ecological Biochemistry* (2nd edn.) pp. 32–65. Academic Press, London. [3rd edn, 1988, pp. 42–81.]

Heinrich, B. (1975). Pollination energetics. *Annu. Rev. Ecol. Syst.*, **6**, 139–70.

Heinrich, B. (1983). Insect foraging energetics. In *Handbook of Experimental Pollination Biology* (ed. C. E. Jones and R. J. Little), pp. 187–214. Van Nostrand Reinhold, New York.

Heslop-Harrison, J. (1987). Pollen germination and pollen-tube growth. *Int. Rev. Cytol.*, **107**, 1–78.

Inouye, D. W. (1983). The ecology of nectar robbing. In *The Biology of Nectaries* (ed. B. Bentley and T. Elias), pp. 153–74. Columbia University Press, New York.

Irvine, A. K. and Armstrong, J. E. (1990). Beetle pollination in tropical forests. In *Reproductive Ecology of Tropical Forests* (ed. K. S. Bawa and

M. Hadley), pp. 135–50. UNESCO, The Parthenon Publishing Company, Paris.

Jain, S. K. (1976). The evolution of inbreeding in plants. *Annu. Rev. Ecol. Syst.*, **7**, 469–95.

Johri, M. and Vasil, K. (1961). Physiology of pollen. *Bot. Rev.*, **27**, 325–81.

Kay, Q. O. N. (1987). The comparative ecology of flowering. *New Phytol.*, **106** (Suppl.), 265–86.

Kevan, P. G. (1978). Floral coloration, its colorimetric analysis and significance in anthecology. In *The Pollination of Flowers by Insects* (ed. A. J. Richards), pp. 51–78 (Linnean Soc. Symp. No. 6). Academic Press, London.

Kevan, P. G. (1983). Floral colors through the insect eye: What they are and what they mean. In *Handbook of Experimental Pollination Biology* (ed. C. E. Jones and R. J. Little), pp. 3–30. Van Nostrand Reinhold, New York.

Kevan, P. G. (1984). Pollination by animals and angiosperm evolution. In *Plant Biosystematics* (ed. W. F. Grant), pp. 272–92. Don Mills, Academic Press, London.

Kevan, P. G. and Baker, H. G. (1984). Insects on flowers. In *Ecological Entomology* (ed. C. B. Huffaker and R. L. Rabb), pp. 608–31. John Wiley, New York.

Kevan, P. G. and Baker, H. G. (1983). Insects as flower visitors and pollinators. *Annu. Rev. Entomol.*, **28**, 407–53.

Levin, D. A. (1971). The origin of reproductive isolating mechanisms in higher plants. *Taxon*, **20**, 91–113.

Lloyd, D. G. and Webb, C. J. (1977). Secondary sex characters in seed plants. *Bot. Rev.*, **43**, 177–216.

Lloyd, D. G. and Webb, C. J. (1986). The avoidance of interference between the presentation of pollen and stigmas in Angiosperms. I. Dichogamy. *NZ J. Bot.*, **25**, 135–62.

Marshall, D. L. and Folson, M. W. (1991). Mate choice in plants: An anatomical to population perspective. *Ann. Rev. Ecol. Syst.*, **22**, 37–63.

Menzel, R. and Backhaus, W. (1989). Color vision honey bees: phenomena and physiological mechanisms. In *Facets of Vision* (ed. D. G. Stavenga and R. C. Hardie), pp. 281–97. Springer-Verlag, Berlin.

Nettancourt, N. de (1984). Incompatibility. In *Encyclopedia of Plant Physiology*, New Series, Vol. 17, pp. 624–36. Springer-Verlag, Berlin.

Niklas, K. J. (1985). The aerodynamics of wind pollination. *Bot. Rev.*, **51**, 328–86.

Opler, P. A. (1983). Nectar production in tropical ecosystem. In *The Biology of Nectories* (ed. B. Bentley and T. Elias), pp. 30–79. Columbia University Press, New York.

O'Toole, Ch. and Raw, A. (1991). *Bees of the World* (pp. 128–42, bees and flowers; pp. 143–53, bees and orchids). Blanford, London.

Ottaviano, E. and Mulcahy, D. (1990). Genetics of angiosperm pollen. *Adv. Gen.* **26**, 1–64.

Peakall, R., Handel, S. N., and Beattie, A. J. (1991). The evidence for, and the importance of, ant pollination. In *Ant–Plant Interaction* (ed. C. R. Huxley and D. F. Cutler), pp. 421–9. Oxford University Press.

Pettit, J. M. (1984). Aspects of flowering and pollination in marine angiosperms. *Oceanog. Mar. Bio. Ann. Rev.*, **22**, 315–42.

Pleasant, J. M. (1983). Structure of plant and pollinator community. In *Handbook of Experimental Pollination Biology* (ed. C. E. Jones and R. J. Little), pp. 375–93. Van Nostrand Reinhold, New York.

Prance, G. T. (1985). The pollination of Amazonian plants. In *Key Environments: Amazonia* (ed. G. T. Prance and T. E. Lovejoy), pp. 166–91. Pergamon Press, Oxford.

Price, P. W. (1975). *Insect Ecology* (pp. 390–406, pollination ecology). John Wiley, New York.

Primack, R. B. (1985). Longevity of individual flowers. *Annu. Rev. Ecol. Syst.*, **16**, 15–37.

Primack, R. B. (1985). Patterns of flowering phenology in communities, populations, individuals and single flowers. In *The Population Structure of Vegetation* (ed. J. White), pp. 571–93. W. Junk, Dordrecht, The Netherlands.

Primack, R. B. (1987). Relationships among flowers, fruits and seeds. *Annu. Rev. Ecol. Syst.*, **18**, 409–30.

Prŷs-Jones, O. and Corbet, S. A. (1987). *Bumble-bees* (pp. 35–52, foraging behaviour). Cambridge University Press, Cambridge.

Pyke, G. H. (1984). Optimal foraging theory: a critical review. *Annu. Rev. Ecol. Syst.*, **15**, 523–76.

Queller, D. C. (1988). Sexual selection in flowering plants. In *Sexual Selection: Testing the Alternatives* (ed. J. W. Bradbury and M. B. Anderrson), pp. 165–79. John Wiley, New York.

Rathcke, B. (1983). Competition and facilitation among plants for pollination. In *Pollination Biology* (ed. L. Real), pp. 305–30. Academic Press, Orlando, Florida.

Rathcke, B. and Lacey, E. P. (1985). Phenological patterns of terrestrial plants. *Annu. Rev. Ecol. Syst.*, **16**, 179–214.

Real, L. (1983). Microbehavior and macrostructure in pollinator–plant interactions. In *Pollination Biology* (ed. L. Real), pp. 287–304. Academic Press, Orlando, Florida.

Rebelo, A. G. and Janman, M. L. (1987). Pollination and community ecology. In *A Preliminary Synthesis of Pollination Biology in the Cape Flora* (ed. A. G. Rebelo), pp. 155–92. Report No. 141. South African National Scientific Programmes, Pretoria.

Regal, P. J. (1982). Pollination by wind and animals: Ecology of geographic patterns. *Annu. Rev. Ecol. Syst.*, **13**, 497–594.

Scogin, R. (1983). Visible floral pigments and pollinators. In *Handbook of Experimental Pollination Biology* (ed. C. E. Jones and R. J. Little), pp. 160–172. Van Nostrand Reinhold, New York.

Seeley, T. D. (1985). *Honeybee Ecology—A Study of Adaptation in Social Life* (pp. 80–106, food collection). Princeton University Press, Princeton, NJ.

Shivanna, K. R. and Heslop-Harrison, J. (1981). Membrane structure and pollen viability. *Ann. Bot.*, **47**, 759–70.

Silberbauer-Gottsberger, I. (1980). Pollination and evolution in palms. *Phyton (Horn)*, **30**, 213–34.

Simpson, B. B. and Neff, J. L. (1983). Evolution and diversity of floral rewards. In *Handbook of Experimental Pollination Biology* (ed. C. E. Jones and R. J. Little), pp. 142–59. Van Nostrand Reinhold, New York.

Stanton, M. L. and Galloway, L. F. (1990). Natural selection and allocation to reproduction in flowering plants. *Lect. Math. Life Sci.* (Am. Math. Soc.), **22**, 1–50.

Stebbins, G. L. (1970). Adaptive radiation in angiosperms. I. Pollination mechanisms. *Annu. Rev. Ecol. Syst.*, **1**, 307–26.

Subba Reddi, C. and Reddi, E. V. B. (1984). Bee-flower interactions and pollination potential. *Proc. Indian Acad. Sci. (Anim. Sci.)*, **93**, 373–90.

Vogel, S. (1983). Ecophysiology of zoophilic pollination. *Encyclopedia of Plant Physiology* (New Series) Vol. 12c, pp. 559–624.

Vogel, S. (1978). Floral ecology—report on the years 1974 (1973) to 1978. *Prog. Bot.* **40**, 453–483. (Ed. H. Ellenberg, K. Esser, H. Merxmüller, E. Schnepf, and H. Zeigler). Springer-Verlag, Berlin.

Vogel, S. and Westercamp, C. (1991). Pollination: An integrating factor of biocoenoses. In *Species Conservation: A Population–Biological Approach* (ed. A. Seitz and V. Loeschchke), pp. 153–70. Birkhäuser, Basel.

Waddington, R. D. (1983). Foraging behaviour of pollinators. In *Pollination Biology* (ed. L. Real), pp. 203–42. Academic Press, Orlando, Florida.

Waddington, K. D. (1987). Nutritional ecology of bees. In *Nutritional Ecology of Insects, Mites, Spiders and Related Invertebrates*. (ed. F. Slansky, Jr. and J. G. Rodrigues), pp. 393–418. John Wiley, New York.

Waddington, K. D. and Heinrich, B. (1981). Patterns of movement and floral choice by foraging bees. In *Foraging Behaviour, Ecological, Ethological and Psychological Approaches*, (ed. A. Kamil and T. Sargent), pp. 215–30. Garland STPM Press, New York.

Waser, N. M. (1983). Competition for pollination and floral character differences among sympatric plant species: A review of evidence. In *Pollination Biology* (ed. L. Real), pp. 277–93. Academic Press, Orlando, Florida.

Waser, N. M. (1983). The adaptive nature of floral traits: Ideas and evidence. In *Pollination Biology* (ed. L. Real), pp. 242–86. Academic Press, Orlando, Florida.

Waser, N. M. and Price, M. V. (1983). Optimal actual outcrossing in plants, and nature of plant-pollinator interaction. In *Handbook of Experimental Pollination Biology* (ed. C. E. Jones and R. J. Little), pp. 341–59. Van Nostrand Reinhold, New York.

Whitehead, D. R. (1983). Wind pollination: Some ecological and evolutionary perspectives. In *Pollination Biology* (ed. L. Real), pp. 97–108. Academic Press, Orlando, Florida.

Williams, N. H. (1982). The biology of orchids and euglossine bees. In *Orchid Biology: Reviews and Perspective II* (ed. J. Arditti), pp. 119–71. Cornell University Press, Ithaca, NY.

Williams, N. H. (1983). Floral fragrances as cues in animal behaviour. In *Handbook of Experimental Pollination Biology* (ed. C. E. Jones and R. J. Little), pp. 50–72. Van Nostrand Reinhold, New York.

Wyatt, R. (1983). Pollinator-plant interactions and the evolution of breeding systems. In *Pollination Biology* (ed. L. Real) pp. 51–95. Academic Press, Orlando, Florida.

Wyatt, R. (1988). Phylogenetic aspects of the evolution of self-pollination. In *Plant Evolutionary Biology* (ed. L. D. Gottlieb and S. K. Jain), pp. 109–32. Chapman & Hall, London and New York.

Zimmerman, M. (1988). Nectar production, flowering phenology, and strategies for pollination. In *Plant Reproductive Strategies* (ed. J. Lovett-Doust and L. Lovett-Doust), pp. 151–78. Oxford University Press, New York.

Suppliers of equipment

General materials and equipment for field work

Ben Meadows Inc., PO Box 2781, Eugene, Oregon 97402, USA.

Ben Meadows Inc., PO Box 80549, Atlanta (Chamblee), Georgia 30366, USA.

Carolina Biological Supply Company, 2700 York Road, Burlington, North Carolina 27215, USA.

Carolina Biological Supply Company, Box 187, Gladstone, Oregon 97027, USA

Forestry Suppliers Inc., 205 W. Rankin St., PO Box 8397, Jackson, Mississippi 39284–8397, USA.

J. Carl Hansen, Lilleringvez 6, 8462 Harlev, Denmark.

LI-COR Inc., 4421 Superior St., Box 4425, Lincoln, Nebraska 68504, USA.

Sciences et Nature, 7 rue des Epinettes, 75015, Paris, France.

Ward's Natural Science Establishment Inc., 5100 West Henrietta Road, PO Box 92912, Rochester, NY 14692–9012, USA.

Ward's Natural Science Establishment Inc., 11850 East Florence Ave., PO Box 2567, Santa Fe Springs, California, USA.

Microcapillaries

Camlab Ltd., Nuffield Road, Cambridge, UK.

Drummond Scientific Company, 445 Sherman Ave., Palo Alto, CA 94306, USA.

Kimble Product, Box 1035, Toledo, OH 43666, USA.

Shandon Southern Products, 112 Chadwick Road, Atmoor, Runcorn, Cheshire WA7 1PR, UK.

Sigma Chemicals Co. Ltd., PO Box 14508, St. Louis, MO 63178, USA.

Hand refractometer

American Optical, Sugar and Eggert Rd., Buffalo, NY 14625, USA.

Bellingham & Stanley Ltd., Polykraft Works, Longfield Road, Turnbridge, Wells, Kent TN2 3EY, UK.

A. Krüss Optronic. Alsterdorferstrasse 220, 2000 Hamburg 60, Germany.
Leica Inc., 111 Deer Lake Rd., Deenfield. Illinois. 60015, USA.

Entomological equipment

Beyrolles, 46 rue du Bac, 75007, Paris, France.
Bioform-HG. Meiser & Co, Bittlemairstr. 4, 8070, Ingolstadt, Donau, Germany.
BioQuip Products Inc., PO Box 61, Santa Monica, California 90406, USA.
Burkard Manufacturing Co. Ltd., Woodcock Hill Industrial Estate, Rickmansworth, Hertfordshire WD3 1PJ, UK
Cabinet d'Entomologie, 43 rue Charles de Gaulle, 49440 Cande, France.
Watkins, Doncaster, PO Box 5, Crasbrook, Kent TN18 5EZ, UK.
Also: Carolina Biological Supply Company; Ward's; Forestry Suppliers Inc.

Pollination bags

Diatex, 16 Chein de Saint Gobain, F-67910, Saint-Fons, France.
PBS International, Easfield Industrial Estate, Scarborough, YO11 3OZ, UK.
Sullivan Company, 250 South Van Ness, San Francisco, California 94103, USA.
Also: Forestry Suppliers Inc.

Tags and labels

Fordingbridge Ltd., Arundel Road, Fontwell, Arundel, West Sussex, BN18 05O, UK.
Joseph Bentley Ltd., Beck Lane, Barrow-on Humber, South Humbershire DN19 7AQ, UK.
Also: Forestry Suppliers Inc.

Fluorescent dyes

Day-Glo Color Corp., Radiant Color Corp., 2800 Radiant Ave., Richmond, California 94804, USA.
Duke Scientific Corp., 445 Sherman Ave., Palo Alto, California 94306, USA.
States Radium Corporation, Kings Highway, Box 409, Hackettstown, New Jersey 07840, USA.
Topline International, Topline House, Bartlow Road, Linton, Cambridge CB1 GL4, UK.

Index

Index